Mothballs

Mothballs

by
Alia Mamdouh

translated by
Peter Theroux

Original text © 1986 Alia Mamdouh
Translation © 1996 Peter Theroux

First English Edition
First published in Arabic as *Habbat al-Naphatalin* by al-Hay'ah al-Masriah al-Amah li-al-Kitab, Cairo, 1986.

Series editor: Fadia Faqir
Literary editor: Georgina Andrewes

ISBN: 1 85964 019 2

British Library Cataloguing-in-Publication Data
A catalogue record for this book is available from the British Library

Cover design by Paul Cooper
Cover illustration by Peter Hay
Typeset by Samantha Abley
Printed in Lebanon

Typeset in 11/13 Adobe Garamond

Published by Garnet Publishing Ltd.,
8 Southern Court, South Street,
Reading, RG1 4QS, UK.

Introduction

༄ঞ্চৎ৯

During the Gulf War of 1991 my place of exile became for me a sad and lonely country. Like so many others I was against the war but powerless to stop it. As I watched the television pictures recording, day after day, the bombs falling on Baghdad, I felt the rift between my image of my own rural homeland and the western city growing bigger. Out of loyalty to the past, to the gardens with their tall palm trees, to the mother's headscarf and to the village, I chose *Mothballs* as the fifth novel in the Arab Women Writers series. Baghdad was partially destroyed not only by the Allies' bombs, but also by change; this novel holds it still in a moment of time now past.

Metaphorical mothballs are sprinkled everywhere in the novel – real incidents, news items, fantasies, dreams, histories, folktales are thrown into a melting pot of reference as the author desperately, frantically attempts to preserve her memories. Mothballs are put in the wardrobe of memory to stop the moth attacking the soul.

At another level this novel reminds us of Maxine Hong Kingston's writing, in which the anthropological element is strong. Alia Mamdouh draws her world with precision. Evocative descriptions of public feasts, weddings, public baths, bazaars and folklore punctuate the narrative. Baghdad as a setting is a powerful presence in the novel. We can visualize the mighty river Tigris, the gritty neighbourhood and the family house. The writing appeals to all the senses as the smells, sounds and textures of a bustling riverside city are vividly called to mind. However, as with Kingston, the attempt to paint a unique civilization and culture with words becomes the attempt to define and preserve self and identity.

Through her rich descriptions and lyrical language Alia Mamdouh creates a 'sweet song for the folk of Baghdad, a song for the places preserved in memory, a song for the childhood and adolescence of an Iraqi woman.'[1] A juxtaposition of the past and present tense and a stealthy change of narration between the first, second and third person unfold the lives of a poor Iraqi family during the 1940s and '50s in the Athamiyyah neighbourhood of Baghdad.

Huda, a fiery and feisty nine-year old, is the central character in the novel. She observes and documents what takes place in the house, the street and the country as a whole by innocently assimilating that which surrounds her. Through Huda we learn of her father the bullying police officer who works at the prison in Karbala, her Syrian mother, her devoted brother Adil, the aunt who waits for a man to marry her and give her legitimacy within this conservative society and the grandmother who is the most potent presence in the novel.

At the core of the book is a household of unfulfilled women. Huda tries to decipher this world of yearnings, frustration and tragedies. Her grandfather dies young leaving his wife to mourn him throughout her life. Her mother, who suffers from tuberculosis, loses her husband to a second wife in another city; the aunt never consummates her marriage while her other aunts are engaged in illicit gay relationships. So the household exists without men, and while much seems to revolve around them the truth is that many of these actions are mere lip-service to cultural traditions. Women are the real holders of the reins and the givers and takers of happiness and approval. Men are propped up by women, the true sovereigns of this poor neighbourhood.

Central to this power is Huda's grandmother. Through her imagination, humour, inherited wisdom and the stories of the prophets that she tells and retells she is created by Mamdouh as a reservoir of past traditions. She is the one with ultimate power to whom all look for love and approval. Huda allies herself with the grandmother, seeking strength from her to defend herself in a society essentially divided between oppressors and oppressed. In Huda Mamdouh succeeds in creating a paradoxical character who is rebellious but humble, sensuous though tough, afraid while still defiant.

Mothballs preserves the past memories, while at the same time exposing the entrenched fear within Iraqi society. When asked to describe herself in one word, Mamdouh said, 'fear'. She calls most of her characters creatures of fear. Through her writing, she explores this dark and deep human emotion in order to control it. 'Fear inhabits the foundations and the walls, is found in cellars and domes, and between keys and locks. It always finds a shelter. No drawer or head is empty of it . . . Our Arab fear is a multi-headed creature . . . I remind myself every morning that fear and death are under my pillow . . . But fear extracted all the pages I have written, as if my books came out of the world to defy torture.'[2] Fear and the darker side of human emotion are implicit in all relations in the book. Huda's initiation into adulthood through her innocent love for Mahmoud exists in contrast to the desperate desires of many of the women in the neighbourhood and their outlet through secret and forbidden affairs. The father tries to cast fear into the hearts of his children and wife with his uniform, whip and pistol. He has an aura of false and fabricated power around him. He, the police officer, keeps his distance from his children and does not allow them to see his emotional side. He remains surrounded by fear and respect, until eventually all his repressed emotions claim and consume him.

The novel also deals subtly with history and politics through Huda's stream of consciousness: the people's adulation for the Egyptian leader Gamal Abdel Nasser, Mahmoud's adoption of communism, the butcher leafleting against the corrupt government and the British, sympathy for the young Hashemite, King Ghazi, and the demonstrations against the British culminating in the 1958 rebellion. The novel touches on various political points but resists engaging overtly in politics.

Alia Mamdouh was criticized for her use of colloquial Iraqi dialect and for not writing a blatantly political novel. Both decisions were conscious. To compose this sweet song for Baghdad, Mamdouh draws on rich oral traditions. In the past Arab women writers used correct standard Arabic language (*Fusha*), which is similar to the language of the Qur'an, to prove their linguistic credentials to puritan Arabists. The *Arab Women Writers* series shows that a clear departure from the use of standard Arabic has taken place. Women writers now use a colloquialized *Fusha* to describe the daily experiences of women. The dominant written Arabic was found to be inadequate to present sexual, religious and social experiences. To be true to women's voices meant that oral tradition had to be brought in. Consequently the dialogues are all in colloquial Iraqi in order to depict neighbourhood life. Mahmoud's awareness of the Arabian Nights, the Qur'an and the folktales is evident throughout the novel, while her ear is sensitive to the way ordinary people talk and to the rhythms and sounds of her home city.

Women in this series create a different language where the patriarch is lampooned and ridiculed, where women's daily experience and oral culture are placed at the epicentre of the current discourse. This series clearly shows how the rejection of standard perceptions about masculine language and feminine language has created a third space within the language from where they can question a culture which is based on exclusion, division and misrepresentation of their religious, sexual and political experiences. They have created a discourse wherein they can gnaw at the foundations of the societies that marginalize them.

Women are treated as a minority in most Arab countries. They feel invisible, misrepresented and reduced. Perceived as second-rate natives, they are subjected to a peculiar kind of home-grown Orientalism. A male native assumes a superior position to women, misrepresents them and in most cases fails to see them. This parallels the Orientalist attitudes with which Westerners have treated the Arab world for so long. Arab women are therefore hidden behind a double-layered veil.

The *Arab Women Writers* series was started to redress the lack of interaction with Arab culture. The series aspires to open a window on the walled garden where Arab women's alternative story is being told. Out of the private space Arab women sing their tales from countries which still, to a great extent, treat them as second class citizens.

Translation is an act of negotiation. 'A translator's job is like buying a carpet in an Oriental bazaar. The merchant asks 100, you offer 10, and eventually you agree on 50.'[3] This delicate balancing act entails being faithful to the spirit of the Arabic text and to its perceived English-speaking audience. The translator becomes something like a double agent, with a sense of split loyalty; the negotiation is yet more exacting in the context of the 'third space' from which these women write.

Peter Theroux's capable hands turned this novel into a song in English. His knowledge of colloquial Iraqi made it possible for him to preserve the local colours in the translation. As a student, he spent one year in Egypt and then spent the five years, 1980–5, as a journalist in Riyadh, Saudi Arabia reporting for UPI, Associated Press and others. So he became familiar with spoken Egyptian and Gulf dialects and the details of everyday life in many Arab countries. Alia Mamdouh's novel, which depends heavily on localisms and colloquialisms, is a challenge to any translator. Peter Theroux's experience and education helped him rise to this challenge and preserve the beauty of the Arabic text. In the absence of many good translators from Arabic into English, a problem partly responsible for the absence of Arab culture from the international arena, his translation is warmly welcomed.

Now I invite the reader to open the book of Arab women's stories. This book is part of a secular project, challenging the foundations of a patriarchal tribal system. It also sets out to challenge Westerners' perception of what Arab women think and feel. If you lift this double-layered veil, you will see the variant, colourful and resilient writings of Arab women, the fresh inner garden. You can hear the clear voices of Arab women singing their survival.

Fadia Faqir
Durham, 1996

1   Latifa al-Zayyat, 'Mothballs', *Noor Quarterly Review*, no. 3, Spring 1995.
2   This quotation is taken from *In the House of Silence: An Anthology of Autobiographical Writings by Arab Women Writers*, Fadia Faqir (ed.), a collection of essays to be published by Garnet Publishing in 1996 to complement this series.
3   Umberto Eco, 'A Rose by Any Other Name' in *Guardian Weekly*, 16 January 1994.

Glossary

ꙮ

abaya: long black cloak covering head and body, worn by women in some Muslim countries.

Abu: title used to address men, whereby they are known by their eldest son's name. Hence Abu Adil is Adil's father.

khishkhash: a mild opiate used to quieten young children.

Umm: title used to address women, whereby they are known by their eldest son's name. Hence Umm Jamil is Jamil's mother.

Reference is made on p. 75 and p. 78 to a precious flask containing hairs. This refers to the 27th day of Ramadan when people visit the Mosque of Abu Hanifa in al-A'dhamiyya and revere the holy relic of the Prophet's hair.

1

⸂⸑◉⸐⸃

The clouds are over your head, and the test is always waiting for you. Just look at your father. It seems to you that he is driving a truck. Your mother is sitting in the back, monopolizing the silence and illness. The rest of the herd are playing inside the detention camp, growling a little, then falling silent.

Your isolated and dispossessed grandmother removed herself from all housework, as if she were created only for worship. She was proud of this distinction of hers, sprinkling water behind us after every meal, with prayers, to obliterate our footprints so that Satan could not envelop us.

She has known devils, your aunt and you. Your grandmother read verse to you from the Qu'ran, to soothe your aunt's hell, and so you will abandon your friendship with evil.

"May God show you the right path and bring you closer to Munir!" she cried to your aunt, and the strength of her faith inspired you.

When Munir thrust his fingers around your aunt's upper arm the mark remained for days, indelible, like that of a slap. He used to show up without notice and leave without excusing himself. When he was silent we knew there was trouble. He chattered about things which could not be understood. He was short and stocky, always wearing a suit and a new tie. His shoes gleamed, and so did his bald head.

He mocked and ridiculed. He laughed and winked. He jumped like a field locust and scurried like the cockroaches in the cesspool. He moved the way the movie heroes did, and pinched me on the cheek when he came in, and slapped my behind when he left. He filled the ashtrays with cigarette butts. He drank a great deal of water and tea.

Yet there was something of an evil spirit about him. You could never tell by his face whether he was serious or joking. He spat on the ground and coughed violently. Your mother vanished out of his way. He always asked about your brother; Adil was afraid of him. I always provoked him, and my grandmother watched everything.

To us he looked big and scary. You learned his age, approximately, when Aunt Najia told your grandmother, "No, dear, he's too old for her. He might be forty, and Farida only came of age a few years ago."

Your grandmother lit two cigarettes and they smoked. This aunt's voice fluctuated between masculinity and femininity. She was full figured and fortyish, and wore gold-framed eyeglasses; her teeth were yellow, big, and jutted forward. She left her hair in two narrow black braids, streaked with white.

She always wore her new long silk robe with the low neckline, embroidered with a tree branch design, over her dress. Underneath it was a pink, beige, or orange slip with the bosom worked in lace and threaded with silver or gold. She would tie a wide headband of shiny black fabric over her wide, high forehead. Her fragrance spread immediately – a fragrance like that of a woman in labour. We gazed at her as if she were a fashionable lady, at the expanse of her wide, freckled chest, exposed to us as, and the crevice between her ample breasts that were pushed up by her brassiere. After we sank into the middle, we were enthralled by the glitter of the big brooch with its coloured stones; the big turquoise stone practically blinded us. When she'd taken the veil from her face in the hallway, she wiped away her sweat with a handkerchief of fresh cloth whose edges were embroidered in brilliant colours, then she buried it in the front of her blouse. She had a long coughing fit and spat out a dense clot of phlegm. Your aunt took her silken cloak with the gold threads hanging from the neck opening down to the waist.

In summer the courtyard was washed. The cushions with the tree branch-patterned covers were arrayed on the straw mats. The brazier gleamed, the coal turned to bright embers, and the teapot and kettle were new. The little cups with their gold-lined saucers and silver spoons were set out in the middle of the round platter, brought out of the old wooden chest. The courtyard was roofed with broad panes of glass seamed with black and grey iron, specked with bird droppings. When it rained, the raindrops reminded you of the devils that went around in your head, and when the sun shone, we did not perish.

Rooms lined the courtyard. Your parents' room was at the end of the corridor, and your aunt's and grandmother's room was at the entrance. It was your room, too – your's and Adil's.

This open courtyard was where you used to receive guests. In winter, the holes in the corners were stuffed with rags dipped in paraffin. We spread out old rugs and worn out carpets, and big pillows with worn linen covers in the four corners.

The grey heaters – the rust – were my mother's job. She took them into the kitchen and began there to clean and polish. She replaced the old wicks. She greased the hard knobs and returned everything clean and polished. She put one in each room, then brewed tea on the big brazier, grilled onions, and warmed yesterday's bread. On cold winter nights we smelled the orange peels as they burned – she used to spread them on the coals to banish the bad smell of the paraffin.

The ground was always the best place for sitting and relaxing. And for dreams and games.

Ahead of us were the steps that led to the high roof, where there were two rooms. The first was spacious and abandoned; there was old furniture in one corner. In the smaller room were heaps of newspapers and books: your and Adil's treasures!

When Aunt Najia came into the courtyard, we knew that the door of secrets had opened before us. When she laughed, our house shook. She shouted, "Huda, wipe off the platter, what's wrong with a little cleanliness?"

You, Adil, and your mother squatted in the corners. Your brother spread out the old newspapers, and pulled the thread off spools to make his kites. He worked like a patient adult; he did not shout or grumble. He was the youngest, the prettiest, the plumpest, the most delicate. You used to divide the world between you and him. He was order, melancholy, and introspection. You were anarchy, insolence, and violence. Your footsteps annoyed the people in the house and the way you walked in the street provoked danger. You were nine years old; Adil was eight. He was possessed by a surpassing ability to bear anguish and pain; you loved to apportion grief, hatred, and love, toward everyone, and among everyone. His head was strong and his eyes were wide, shining with sects and minorities. Their colour at night concealed the echo of sensitivity; in daylight, lights mingled in their honey-coloured waves. He was beautiful – venerable. He stood before you and you looked at him. You gave him titles – you barked at him. The day passed, and another, and another. You knew that his beauty was your greatest joy. The lines of his face, his nose, his lips, his silence, his backbone. The shadows of his casual sympathy, the depths of the resistance which people like you never know. All of this turned you against him. Siblings' fears are not written down or publicized. They proceed, step by step. You lit all the lights so that he did not walk alone. You were with him; no, he was with you. He was the one everyone loved, he was the one who loved you. It was you who pushed him towards the wheels of the wagons drawn by decrepit horses. It was you who were frightened by the creeping of the horses, though he was silent. You cried "God is great!" in the street, you shrieked. You took him behind the graveyard to frighten him. You dressed him in your father's uniform and saluted him. You got him embroiled in wild dreams. You were sure of nothing but his regal face. Adil's face was created as if he were meant to return to Heaven young. You used to take him up to the roof, place him at the top of the steps and push him down. He did not raise his voice, cry, or whisper. He did not give away this secret. You were burdened by the ways you tortured him. You were an expert at capturing him and hiding him. You stole the money that his father gave him, sweets, apricot paste, dried apricots, and the bunches of wrinkled figs that your grandmother put aside for him alone. He never protested, he gave you things in front of everyone and away from everyone – when they were asleep, when they were out, when they returned. He loved you as if you were the last sister in the world. He knelt before you, gave you his portions, made winged animals for you, frightening bears, gentle toys, and did not bother with talk. You imagined him standing up, his chest ready to receive bullets. He would close his eyes, his tears would flow, his pulse would stop, he would not raise his spindly but soft arms in the air and say "No!"

Since that time you have been alone, sinking into infernal stoicism. He stood in the doorway, defending you from their hands and feet, the whip, and shoes. He cried instead of you, and your rage mounted. You have brought all these curses upon yourself yet you always found someone to blame.

Your mother moved as if she were climbing a high mountain. She brought tea and biscuits on a wide, flat tray. She offered each person a fan. She sullied no one with her voice, responding to Aunt Farida's shrieks and shouts with a brief nod of her head. She gave them Adil and Huda – what more do they want from her?

She was extraordinarily slender, fair-skinned and tall. Her hair was the brown of an old walnut; her eyes were honey-coloured but showed no light. The skin of her face was dry, her cheeks hollow, her teeth crooked. When she laughed, she asked God's protection from Satan, and her facial features became tense as she remembered that laughter is a sort of sin.

After pouring and serving the tea, she sat on the low wooden bench like a dejected sentry. She opened and closed, rinsed and dried, came and went. She finished everything slowly: cooking, eating, loving her husband.

Her sharp coughing travelled through walls and windows. You heard your grandmother reading prayers to her, your aunt as she cursed her. You and Adil were encircled, by that cough. You were moved to your grandmother's room, because of fear . . .

We did not know. We did not understand. We did not want to know. And they did not want us to know.

We called her "Mama" only when we were frightened or needed help, but after giving birth to us she lowered a dark curtain of secrecy around her narrow domain.

She took all her guidance from your father: she did not dare refuse. She walked toward the supernatural with her ailing chest with the same noble forehead, dreaming of truffles and the nourishment of her cheeks and thighs in case the inevitable faltered. Your grandmother loved her. And she hated no one.

Aunt Najia began to pace in front of your grandmother. She joked with her, she chatted her up, and teased her with grand titles. This aunt made no distinction between any of the women. When your grandmother appeared, she wanted her to herself, and when your father's sister came, she got her into new positions, saying things she never thought of saying before. When a changing rapture appeared on the horizon, she embraced it fully, never remembering what had gone before: the past, apprehensions, the first stammering. Your grandmother now admired her; this was what she wanted in her possession. She sprinkled flame on her and emerged from all discords. She fell with her whole height and width as if struck by an all-consuming itch. Her voice vanished and the dripping of blood was heard. Her long robe was drawn back a little from her taut thighs, and legs as long and slender as a goat's. She kicked off her pointed sandals and they landed across the room. She rolled up the sleeve of the robe, higher, higher, up to her armpit. This was where the smell of steam and sweat came from open pores and the extending folds of her limp forearm. The hair between her armpit and forearm was long and black. When she was intoxicated, she undid the brooch and it dropped on the floor. Aunt Farida sat across from her with her legs open, and your grandmother asked God's forgiveness:

"There is no strength or power save in God!"

The words detour: "Even God's words sound good coming from you."

"Listen, Najia, God help you."

"Oh, even my name sounds good coming from you."

"Listen, you know I don't like that kind of talk."

"Fine, fine. Don't get upset. By the way, is Bahija Khan going to drop in?"

Their voices rose, your head bowed to that remote depth and you chose your first relation between the edges of the mysterious split: the enigmatic body.

You gave off strange vibrations, you don't know how, or where they went, or who would feel them.

You will never see these strange women again in your life. You love listening to them. They are gold mines: if you go and extract it, the sun will shine; if you leave them in the belly of the earth, the belly will split asunder and produce a different posterity.

Were these the corrupt women you have heard about?

Women: souls painted with fire, bodies over which the open air passes, making them radiant, over which the salts of the sea pass, making them blaze, at whom fear fires its incomparable rays.

They screamed at you; they watched you.

You were there, in that courtyard, listening to horns that bleated at the threshold of your soul. The neighbourhood in which your body lived was agitated. You did not retreat. They dragged you, gently at first; they beat you, and your mother went into the distant kitchen. Her shame was striking in its candour, among those brilliant souls.

Aunt Najia started shouting again. Her voice had become riper than any other voice you'd heard in your life:

"I want Bahija Khan."

They gave the title *Khan* to proper ladies between the Baghdadi period and the Ottoman occupation. The father, grandfather, or brother drove them into isolation and degradation, so the women took both the title and the abandonment.

Bahija was your grandmother's younger sister, the daughter of her stepmother.

Beautiful, plump, tall and broad, proud and haughty, she was about to turn thirty. All the women you knew plot against her. If she stepped into Aunt Najia's trap, it was because she resembled her. If she went to another, it was because that was her nature.

You did not realize all this. What was occurring in front of you left its mark, a step and who knows where it is going to lead you. That whole network of arms and legs met unwritten covenants and invisible charters. What went from this to that was bound as a kind of love from whose shadow there was no escape.

Your aunt's voice emerged sharply from her throat: "I want you to go like lightning to your grandfather's house and tell my Aunt Bahija to come quickly."

2

⤚◎◎◎⤙

The big house was one kilometre away from our house. Our grandmother's sisters lived there, and the widow who had suddenly grown
senile after the death of her wealthy husband, leaving behind Bahija
and Zubaida and Nahida. Nahida had two daughters younger than
me, and two sons older. Zubaida was barren, and Bahija loved women.
When I entered, everyone looked at me with constant superiority and
turned their heads the other way when I passed. We called this the
house of dreams. We wore our finest clothes when we went there. My
aunt combed my hair and pinched me on the arm, saying:

"I swear to God, if you break anything over there, I'll kill you."

We tasted all sorts of fruits there, fresh meat, and exotic types of
sweets and sugary pastries, liberally given to us by Aunt Bahija.

When you saw yourself in the street, your fire was stoked. There
you flung yourself into the tumult and different ways. You stood in
front of the vendors, shooing the flies away from the white cheese
wrapped in fresh palm leaves. You greeted the cheese seller, Abu
Mahmoud: "Hello, dear Abu Mahmoud," and stole a fresh cucumber

and a date whose sweetness burned your mouth. You did not look up. The alleys of your neighbourhood were filthy, littered with onion and aubergine peels, okra tops and fragments of rotten bread, remnants of black tea – all of it took you by surprise. You slapped cross-eyed Hashim, the son of Razzuqi the carpenter, called him a name, and ran away.

"Hey, still cross-eyed?"

He ran after you, the edge of his dishdasha, his ankle-length shirt, in his teeth, his feet trampling through the mud and garbage, and then slipped and fell, and everyone laughed.

You ran and jumped over the gutters and the children. You ran through the house of Mrs Rasmiya, the neighbourhood nurse. Her door was always open; a stained white curtain with holes in it hung in the doorway. You heard the voice of her husband as he beat her and snatched the proceeds from the injections she has administered, and he laughed as he bumped into you: "Hello, Huda. Give my regards to your father."

The houses of Baghdad had stone steps on the outside. You loved standing on these steps, getting to know the herd that waited and knew how to stand. You stood on one of them one day and said to Mahmoud, the son of the cheese seller:

"Look, I am as tall as you are."

There the Baghdadi women sat or lay on cushions, old carpets, and straw mats. They all held fans, and their veils covered only their heads. Their nightgowns gave off the smell of onions and parsley, eggs and sweat. They opened them a little, as if opening their souls. When a stranger walked by, they exchanged glances among themselves and covered themselves until he had passed.

The doors to the courtyard were made of old blotchy wood whose coats of paint were peeling in every corner. When winter attacked, everyone waited for Abu Masoud, the painter. In the middle of the doors were shiny or rusty hand-shaped iron knockers. We stood before them, banged the knockers, and ran away into other streets, far away. We raced and gradually became familiar with this district of houses where no one complained of hunger, which were tidy, tall, spacious, surrounded by towering trees and unfamiliar flowers, and built of gorgeously coloured or painted bricks. The girls here wore wide, pleated skirts and short-sleeved blouses, with coloured ribbons in their hair or around their braids. Their hair was always combed, and their faces freshly scrubbed. Their skin was clear and radiant; their blood sang with health. Their food was fresh meat from Mr Hubi the butcher. Hubi was about forty, fat and red-faced, with a big belly and a broad, slow voice. He butchered lambs, singing, as if he were watering his garden.

This was the only man whose orders were obeyed. Everyone in our street wanted to find room in his shop. Even the dogs and cats bathed in the smell of his tender, freshly killed meats.

Cows, calves, and lambs hung there, bathed in blood and soothed with verses from the Qu'ran.

Anyone who stopped in front of his shop would be greeted with every compliment and blessing that came into his head.

Hubi knew everyone: the family trees of the people who lived in the palaces far away, overlooking the Tigris and the old wooden bridge, the pedigrees of the houses that ate their meat in silence, the histories of those who ate bones and broth, and those who threw meat to their dogs or into the garbage.

To us, Hubi was more important than the king. The King of Iraq was young. A portrait of him with his uncle hung in Hubi's shop, surrounded by spatters of blood and animal remains. Hubi sold meat only in the afternoon: the morning was for slaughtering and butchering. He sold the hides and heads to Abu Mahmoud, and the sheep's livers and testicles to restaurants. His shop produced everything: problems, quarrels, and even secret leaflets.

In the afternoon, our neighbourhood in al-A'dhamiyya came to a stop. The noon, afternoon, and sundown prayers were called from the ancient Abu Hanifa mosque. Faces came, figures passed by, and arms strained. Hubi sliced away the shanks and legs, intestines and shoulders as if he had been created a butcher at birth.

The day your grandmother sent you to him, you raised your head to hers.

"Huda my girl, don't lose the money, or we won't have meat for a week."

She said no more and you were on your way down the street, absorbed by the jingling of the twenty fils coins. At that very moment you could have flown to the next street. Buy candy floss, colourful lollipops, and currants. Fill your hands and your empty pockets, your reckless head, and your tongue dry with all the forbidden things you had seen only in the hands of the children in these other streets.

Steal and lie. Argue and make up excuses, for in Baghdad people take opposite paths: if you steal, your corpse will not be laid open, and if you lie, God is forgiving and merciful.

That is what your grandmother taught you, who stood before her prayer carpet all the time, and between times, in heat, cold, and rain. Her only passion was for God. She explored herself with prayers that never ended; she drowned everyone in supplications, and divulged no secret. She contrived no tricks, she stirred up no scandals, or played with anyone's nerves. She stood in the courtyard or on the roof, saying:

"Lord, bind me unto you, and never let me close an eyelid without a thought of you, O most merciful God."

She used her imagination, wit and wisdom and the stories of the prophets shone as she put us – Adil and I – on her lap. She came to the tale of the prophet Joseph. She spent a long time on this prophet, describing him in a reverent voice: "My dear, it was he who was the death of Potipher's wife."

You asked her: "Who was Potipher's wife?"

"He stood alone against that treacherous woman and his accursed brothers. She was like Lucifer himself, but Joseph pushed her away. Later on he got an inspiration from Almighty God."

Adil's voice: "Who did our lord Joseph look like?"

"No one looks like him. I don't know anyone he looks like."

She wasted no words. She freed herself from all her difficulties, by referring them up to the Omnipotent Deity. You learned about the first of the devils – Potipher's wife – from her. That wanderer, seduced and exposed, became my premonition. There my gaze fell on her for the first time. I saw her, named her, and compared her with the other women, dividing up what she had among all: your aunt Farida, your mother's and father's sisters. But what remained was still abundant.

The day I read the Qu'ran, I read the Sura of Joseph. It opened before me new territories for questions and battles.

With one blow I tore up all the tombstones, as if going into darkness with everyone.

You continued seeking lazy mornings when you did not have to go to school, for vast distances in which you exhaust your anger and love of questions. You did what you were asked in a different way. You wished you had Mahmoud's muscles, Hubi's fingers, and your father's legs. Your head was dizzy from being hit. They had stuffed it with slaps and commandments. You took the remains of sins and stayed that way, peeping through the cracks in the doors and window panes at Potipher's first wife – Aunt Farida – and your mother's two sisters. They were in one another's laps, their flesh trembling, liquids and lapses coming from their lips, new truths from their heads.

When your grandmother went up to the high roof, she took her prayer carpet in one hand and the Qu'ran in the other. She murmured prayers to protect us all. She did not look down or turn around. At that time the two beloved aunts came in and went down. They left their arms linked, their cloaks slipping from long famine; the whole of the room enters into the madness, sighs, appointments, streets, and people. Drops flew off them: "Kill me, dear Bahija, kill me, my dear."

Some went this way, as if it were fate. If your grandmother stayed away, it was because this match was between her and another entity: her soul. And if your mother was absent, it was because she accepted her only good fortune: your father.

But these women were descending to Paradise, not waiting for a lift or contrary sura. They were arm in arm and heel to heel. The flowing channels of the body freed things that were pent up. Aunt Farida stood behind these boundaries, waiting, ready, trained to be ready. Her training was done from the onset of first awareness; when the hour came, they did not delay a second. Women, women, were all around you, recorded in the maps of cities, desired until the Last Judgement; they grew, they proceeded slowly, marching, listening to rumours, walking carefully around the forbidden area: men.

Whenever she crossed a step, the soul went out of the circle and swept away with it this sedate, divided universe, the man and the woman, the boys and the girls, into thousands of pieces and thousands of sighs.

When you confronted your aunt, face to face, she took you with her to the street for visits. Your hand was in hers, and her black cloak defined her new and mysterious figure as she passed by the shops and the coffee-houses. Her swinging walk in her high heeled slippers filled their heads with fantasies. She slowed down and paused, walking as if dancing. The children moved aside slightly so that we could pass. The young men let out low whistles and the men sighed with admiration, but she looked at no one. When we are well beyond them, they call out:

"You're killing me!"

I hid inside my old clothes and my excessive thinness. My rage mounted and fell on me, making my head hurt, so I hurt my aunt's hand. I slipped away from her, and walked far in front of her, not looking back.

Look and stop there: for after your two aunts left the room, Aunt Farida made sure her voracity will begin. She went in and turned on the light, smelled the odours, looked at the ground, and handled everything slowly. She approached the cushions slowly. Everything was tidy and orderly. She leaned over and inspected it carefully. She opened the chest of her body and spread it on the floor and the trance began.

Your aunt left school after elementary level. She sat in the house waiting for Mr Munir. Your father gave her a monthly allowance, and your grandmother too.

From there she rotated between the big bath and the neighbours' houses and your grandfather's house. At these places her physical vitality bulged, and any disruption in this meant slow death or a huge scandal.

Mr Munir, patient, blessed, a cousin, old, rich, unemployed, ugly, hesitated to propose to her, but when he came he found her suitable, an existing oasis, and attractive as well.

Your aunt was the most beautiful woman in the house and the whole neighbourhood. She wanted to use that beauty to break some of the padlocks. The fear for her was compounded when she applied kohl to her eyelids. Her body was in excellent health; her round thighs swelled and floated in the tight clothes selected by Rachel, the Jewish seamstress. Her hips were high, her legs full and her bosom taut. Heavy and erect, her neck was long, and she had high cheekbones like your father. On her lower lip, a small mole, which we call a

"Baghdad mark," made her even more alluring. A beauty spot on her left cheek convinces all men and women of the power of the desire within her. Her eyes, sculpted with a sharp chisel, were black and almond-shaped, and her eyebrows were thick and rarely trimmed.

Her face changed, from the haughtiness of a princess to the remorse of an adulteress. Her hair, coal-black, was now invaded by white wisps. When she laughed, you could see her dimples. When she was silent, the air she gave off had a hoarse hiss. Your aunt sang, too. Her voice, singing folk songs or mad love songs, climbed and flashed with everlasting Iraqi pain. She lowered herself to the ground, sniffing the delightful smell of grilling meat. She loved the smell of other people's bodies, their sweat and pungency. Her tongue wandered before departing; her mouth was dry. The bones of her chest tingled. The arteries in her thighs sang, and she was penetrated by thousands of passions; she trembled and her fingers reached down to her belly. She felt within herself, within the circle of her region – secret sex but tumultuous sex. She wet her lips with saliva, breathing one heavy sigh after another, looking down at her rosy nipples, her pores open, crushing the ribs of this frenzy. She undulated like a firehose, not covering herself. You were behind the window watching her. Her hair was loose like a gypsy's. No one called you away; no one paid attention to you. Your two aunts were content, one of them sprawled on the mat, her face turned to the glass ceiling, the other rolling a fresh cigarette, sealing it with saliva and lighting two cigarettes. She offered one to the sleeping woman. They took the first puff, and Aunt Najiya coughed. "I only like Ghazi cigarettes." Cough. "I don't know why I listen to you."

Your grandmother fell forward onto her fingertips on the roof. The voice of the muezzin called the evening prayer, and the two aunts roused themselves to go to the bath, and after washing they stood with my grandmother to pray. The whole neighbourhood was transfixed in awe. Grandmother sought protection from Satan. Her breathing was quick with supplications, the holy names of God. For the first time, Adil's voice rang out: "I'm going up to fly a kite." Your mother sought refuge in her room, Aunt Farida pulled her clothing over her thighs, and Grandmother stood before all: "Lord, forgive us in this world and the next."

Adil was already on the roof. This was Wednesday, and your father comes on Thursday.

3

‿ဖခုဖ

Thursday was the day of the market bath. Your mother prepared a bundle of clothes for you: a clean vest and an old dress, unbleached cotton, elastic, dark ribbons for your hair, the comb with the wide, broken teeth, your open-toed sandals, her seamstress's horse blanket, ribbed its length and width and lined with cotton, a folded shawl decorated with little circles and squares, a loofah, and soap. You tied up the bundle and stood in front of it.

This was Aunt Farida's day. She took her hanging straw bag and filled it with a bottle of water, some pears, a small melon, a pumice stone, homemade soap, a black bag, her blue perfume bottle, clean clothes, and a cake of cardamom soap.

Your grandmother, whose asthma had troubled her lately, your mother, ill in her chest, and Adil, who had grown up a little, all stay at home. The bath in the ancient house was old and broken down, it was still being repaired. Your father painted it the first time, replaced the old punctured barrel, and paved the floor with new brown cement. He went into it before anyone else, and your grandmother was the last to leave it.

Your aunt was the only one to frequent the market bath. The taste of the journey from the house to the bath, walking through the alleys, calling out to friends encountered by chance, scrutinizing new faces, and before this, leaving the house. We spent the whole day there. We boiled eggs and potatoes, fried kebabs, and grilled onions, then covered the food with flat, warm loaves of bread, and packed it all into paper bags. The day Thursday arrived, I held my breath, my skin peeled there, and my blood ran clear. There I was devoured by the muscles of my aunts, the sisters of my father and mother: Najia, Farida, one-eyed La'iqa, and Umm Suturi, opening their layers of pores and putting me in the trap. I stumbled about amidst the tons of flesh and breasts, bellies, and buttocks.

The bath in the Safina district was far from us, in the other neighbourhood. We went through alleys and emerged in streets. We turned to the right and then to the left, and from the beginning of the street came the smells of women and children, mothers and grandmothers. Their cloaks fluttered, they were blooming and alert, their cheeks were flushed, and no matter where you looked they all busily chewed gum. Women came and went. Their heads were tense, their feet were blistered, and their nail polish was cracked. And you could scarcely hear their voices.

In front of the great door, painted a dark grey, the boys played marbles. Black wooden benches were set in the four corners. Warm breezes blew from inside, and a tall woman in her fifties, slender and ugly, was standing in front of a wooden partition. Her chest was bare, and her breasts were like two withered pears. A damp shawl was pulled around her middle. Her hair was long and hung in her face. She was shouting at everyone.

"You want someone to rub your back or not? Put your things down! How many of you are there? Five? Five costs thirty fils."

Your aunt stripped off your clothes; she was in her underclothes. She looked to the right and the left. Your aunts came in, one after the other, and languidly undressed. Everyone looked at everyone else. You saw everything here seized by the fever of these features: eyes without kohl, cheeks without ceruse, slack lips, and yet flawless bodies. Skulls and bones.

The broad metres of the bath became a source of play and activity.

The first place was not very warm. Children and women dried their hair and limbs. A commotion of Iraqi exclamation, the drone of aged women talking. Women massaged one another. When we went into the second room, the clouds of vapour were rising. My Aunt Najia's voice:

"Listen, Farida, I can't walk inside. I can't catch my breath – I can't breathe. We're better off staying here." Aunt La'iqa answered: "Go on ahead. As soon as you're there you feel numb. The steam will absorb the cold and damp."

Umm Suturi walked ahead of everyone. She knew the way, and she knew everyone: Aunt Najia's neighbour, Aunt La'iqa's friend, the neighbourhood's cheap seamstress. She sewed men's dishdashas and pyjamas, which they bought for circumcisions, funerals, weddings – to trill and clap.

Aunt Farida did not know what to decide. She was the youngest of all, eighteen years old. The women's eyes scanned her body attentively.

"Where is Huda? Come here – even in this fire you'll make friends!"

There you saw the whisper of skin soaked with steam, water, and perspiration. The smell of armpits and buttocks, of urine, mutters and grunts escaping their lips, and shouts across the water barrel.

Everything passed before you: hands took you and cuddled you between their legs, calling the names of everyone you know, undoing your braids. You were showered and soaked, and bowls of hot water were poured over you, on your head, over your delicate frame. You wailed: from there you sent the first speech recorded with anger, you cursed, paused, sniffed, paused, and asked.

You looked with loathing at all these details. Women, all naked, as if they had just been raped or tortured. They laid old towels over the low wooden stools and squatted on them. The floor of the bath was as hot as a grill, and they cried out to one another and shrieked, and brawled with one another. There were no partitions in Iraqi baths, the borders were open, and the one language in which everyone conversed was physical touch. As if they had all been detained beyond the sky and today they had descended to the floor of the bath. There I made my first discoveries and won my first arguments; and shouted "No, no" among the long "Yeses" you hear from everyone else. Only there you were given the bloody title of Huda, a flaming fire.

I slipped away from them all, glided between their legs, and the cakes of soap pushed me far, and I landed in the lap of one woman, her face covered with soap lather. She shrieked, "God Almighty, God damn you and damn the bloody day you were born!"

I hid the cakes of soap in the big buckets, dunked the bowls into the hot water and poured it over their heads, burning their scalps and skin. I pissed in the great tank. I clamped some of the children between my legs; I kept this one away from that one and began to massage their heads with the pumice stone until they were bloody. Before the huge quantities of clean, hot water I observed my first innocence and united with it, drew on it with a pencil and confessed: as if I were saved from the flood today.

My resistance ripened on the oil fire and the wood logs, blazing and transforming into a creature I have just come to know; Huda, covered with sin, affliction, and ruin, was dragged like an animal to complete the first blessing; and after I am left for a short while between the waters; the offspring of Iraqi women reach the perfection of their beauty.

Umm Suturi emptied the bowls of hot water over my head, and soap went from hand to hand among my aunts. They rubbed and twisted my braids. I died among these women's fingers; my eyes were blinded by the soap lather. Aunt Najiya clutched my thigh as if she were holding a chicken leg. My aunt sighed and leaned over her knee, her breasts putting me into a stupor. The soap, steam, and all that noise; I was an egg thrown on to the ocean. I was moved from one lap to another and I see.

There, crying, wailing, and kicking are useless. After a round of washing, you were left alone and free. They stuck their tongues in your ear and sucked out the water left there. They braided your hair into ponytails, and you watched them all. The steam at the end got into your eyes, ears, and mouths.

Laugh and look well: the hair on the limbs is delicate, fine, coarse, long, short, plucked out. Then all these limbs descended at once and removed their underclothes. You gaped at that continent of femininity. The black bag sewn with big stitches in white thread first appeared on their backs. Every woman turned her back to her neighbour, and every one who let down more coils of dirt than the other proved her strength and youth.

You turned around with them when they stood. Their height blocked the walls, which were spattered with waterdrops. Sweat stimulated the appetite to drink water and eat fruit. The talk was of neighbours, children, and husbands. Rachel, the Jewess, whose second son was aborted at the hand of Rasmiya – the "needle lady", the midwife. There were no great scandals in our street, nor any great abominations in our houses. The men intensified their glands in obedience to women, and the women waited for their husbands on the benches, on the iron beds, on the ground, on high roofs, half asleep, half dead, half . . . half.

Your aunt hurried behind you. She wanted you to stand in front of her:

"I swear to God I'll kill you, may God take you and give me a break!"

Aunt Najiya answered her: "Come here – I'll finish washing you."

I slowed down, and stood among them. All the vapours and odours made me dizzy. Aunt Najiya, standing near me, released a fart. I raised my head toward her and laughed loudly. Suddenly she struck me on the face with the bag:

"Laugh, you impudent thing. Just wait, I'll teach you."

She lifted me up as if I were as light as a punctured ball. I was squeezed by her arms; she began scrubbing my forearm, panting, "Why do you make me smack you? Aren't you afraid of anyone? God Almighty. Don't you get tired?" She slid down to my belly and thigh. "She's weak, like her mother, like she's eating on credit."

She curled me up between her thighs. Her hair was loose, long, and wet, sparse and fine. She did not see very well; her eyelashes had fallen out, and her eyelids were swollen.

You gave in and slept. Your skin was now vacant, emptied of its secrets; filth too was a secret. Thus far death had not come to any of you. Until you were nine you did not know what death meant to you.

All the people you knew and loved were alive, in front of you: your brother, your mother, your father, your grandmother, and the neighbours' children. Mahmoud, who moved to middle school, you used to call him Mahmoud Snotnose. He used to chase you and try to hit you, and when you ended up face to face you laughed at him, and he wiped his nose with the hem of his dishdasha. The mothers of your friends were still alive, and their fathers too. You did not know what death would do if it came.

On religious holidays, you all went to the cemetery behind the mosque. You visited the grave of your great grandfather. Your grandmother stood before the grave; she did not cry, nor did she wail or smite herself in grief. She murmured verses from the Qu'ran, her voice hovering over the dust.

She read aloud, and her voice rang out, painfully sharp. It floated over the expanse of the cemetery, moving the women to sob. You used to watch her as she filled your head with the dark side of death, as if she were opening up all the holes in all the heads, land, and souls. There she used to exercise, standing at her medium height, her slenderness, her clean cloak, her heavenly face: how did the wing of life droop to death?

When your aunt called to you, "God take you," she did not go into details. "Take you" perhaps pushes you beyond death, and you begin to ascend. Your height, the muscles of your thighs strengthened, and your chest began to thump from within – your heart, too, wanted to ascend.

You did not know what had happened to you. You saw yourself on a wooden bench in the huge, cold dressing room, Umm Suturi was over your head blowing warm foul breath on you from her big mouth, her thick lips murmuring a few verses from the Qu'ran. You knew it was the Sura of Ya Sin which you knew by heart. She kept breathing on you and started to pull your hair, smacking you gently and rapidly on the temples. She rubbed your chest, and the flesh of her creased belly which touched your belly, leaving her lower half tightly wrapped in a sarong of delicate Indian material. Her soft breasts brushed against your inflamed cheeks.

She dressed you quickly, squeezed your hair dry, draped you in large towels and put another under your head. You stayed that way until everyone left. You slept like the dead. From there, you conjured up the bodies, the thighs, breasts, braids, basins of hot water, and the soap lather. You entered all of them in that hell and began your first resurrection. You invited them to shriek at one another, to leap about, to be consumed by fire. Their voices cried to the heavens. You opened no window for them, you read them no sura of the Qu'ran. That was your place. You became sovereignty in all its magnificence and power. You did not intervene or even appear; you did not threaten or menace. You let them plunge into one another. You cut off the electric current, you scattered snakes in the baths, tore the clothes out of their shiny bags, and smeared them with mud or buried them in cesspools. There, my virginity shone. When I reached this point in my sleep, Aunt Farida was by my head, her complexion peach-hued, her nose shiny, her eyebrows drawn in kohl, her eyes exploring the sleeping girl. She sighed and gasped, leaving me to sit nearby with a white towel around her bosom and hanging to her thighs. Her head was tense, and I did not move – it was as if I was nailed down. I opened and closed my eyes, looking at the water drops on her silken back. I swallowed. Aunt La'iqa went over to Aunt Farida, plump and flabby, her belly like a barrel and her thighs rubbed smooth with fat: her skin was a waxy yellow colour, and the hair on her limbs was blonde. She did not cover her body: "I'm dying of thirst. Where is the water?" She leaned over and took out a bottle of water and some pears. Umm Suturi and Aunt Najiya were in front of me. The voice of that aunt shrieked in my ear: "Look at this poor animal – she's still asleep! I hope she never wakes up!"

She looked aside at Aunt Farida, who had begun to put her clothes on: "God guide her. She's still young."

"No," was Aunt La'iqa's answer. "She has been impossible from the day she was born. Remember when we were giving her *khishkhash* and she wasn't yet forty days old. God help us when she comes of age! Marry her off quickly, before she disgraces us."

Amidst the steam and the sounds of drinking Umm Suturi's gruff voice sounded:

"And who would marry her? She's weak and pale. She's skin and bones. Look, Farida, I'm afraid she has her mother's illness. What about examining her?"

They examined me. One day Mahmoud said to me, with a street light separating us, at the top of our street, "Your mother has tuberculosis."

I chased him, a stone in my hand. He did not run away from me like the other boys. He stood there. I held the stone in my hand, my face a fountain of flame:

"Son of a bitch!"

He did not disappear from my path. We stood together, face to face. I was smaller than he. I was a female and he was a male. It was I who chased him – something he was unable to do. No, but he was able to do many things: run, play, escape from my father's face when he saw him in the street. He taught me arithmetic with his sister Firdous. The first time, I stood and dropped the stone and asked him, "What is consumption?"

"I don't know. My mother says her chest is pierced with holes like a sieve."

I bowed my head, then raised it. "Maybe everybody's chest has holes."

"No, just your mother's. My mother says, 'Don't play with Huda – she'll infect you.'"

Infection, tuberculosis, isolation! I wanted to raise my head again in front of Mahmoud, but was unable to. He was the bravest child in the neighbourhood. I chose him for myself. This would be my first man. That is how free I was throughout all those years. We spat on the ground and looked at our spittle – was there any line of blood? And when we saw nothing, we shouted and screamed and ran through the streets, we hit some people and made jokes with others, we pulled off women's cloaks and knocked men's hats off, and knocked on the front doors of houses and ran away.

He was always saying, "My mother says 'All the girls in the neighbourhood are like your sisters,' but you're nothing like Firdous. She's sensible and you're like the Devil!"

"Are you afraid of the Devil?"

"No."

"Listen. Do you like Hell or not?"

Since that time Mahmoud kept his nose clean. He changed his long dishdasha once a week. He wore sandals, and the fair skin of his face grew red and sweaty from playing, jumping, and running. We played from three o'clock until five in the afternoon. We went into our houses, drank water, peed, and then went back out to the street.

The girls played "hide the beads." We made piles of dirt and sprayed them with water to make little houses in which we hid the coloured beads we had stolen from our grandmothers and fathers, red and yellow, blue and black beads. A few metres away, our voices split the air: "Huda, you're cheating!"

My success in the street was a form of cheating. I usually guessed the number of beads buried in the mud so I took all the girls' beads. I put them in the bag I had tied around my waist. Winning put me in front. Mahmoud on the other side played with a top and won – his top turned, and turned, and turned, as if it would never stop. It never tipped, it never shook, and he pulled the string tightly before whipping it out on level ground. We all stood to watch, while others did as he had done: Suturi, Nizar, Hashim and Adil too. We watched and shouted to one another. We sang to Mahmoud's top, chanting for it not to stop, and mocked the other boys' tops. Our blood was up – our shouts nearly broke the neighbours' windows. We acted like lunatics. Adil and Firdous were with me.

"Oh, God, don't let his top stop, Oh God!"

Mahmoud's top stopped when my father appeared.

Every two weeks my father left for Karbala on the dawn train, and came back to us in the afternoon. His shadow, his name, and his voice went right through us. We huddled together like terrified puppies. It was no use burying our heads under a pillow or wriggling up against our grandmother – he could hear our pulse as soon as he entered our street, and we could hear him muttering between his teeth – we were about to drop to the ground. He carried a small, old valise, the colour of an old boot. Everyone greeted him, standing up as he passed by, utterly quiet. The police officer's emblem entered the street in silence and anticipation. A pistol hung in a holster at his belt, between his waist and thigh, a tool that did its work in the house and in the neighbourhood; with it he killed our repose; through it he became complete, generating terror and respect, thus disposing of his anxiety and sense of struggle.

When he passed, women opened their cloaks so he could see their bodies undulating and their winking eyes, and their teeth poised on their lips. Men tightened their belts over their blue, white, or striped dishdashas. They adjusted their headgear, determined to stand up and greet him. He passed, looking only straight ahead.

He wore the high woollen *sidara*, which was similar to a garrison cap, on his head and a single star on his shoulder. His khaki uniform showed off his slimness and height, and his high black boots were always shiny – he avoided the rubbish and puddles of filth. He walked like a peacock. He was graceful and good-looking, and brown-skinned, and his brown eyes were wide, piercing and reckless. His nose was long and high like the noses of the fathers in the other streets. His lips were narrow and prim and always red. He had high cheekbones, and his hair was the colour of old silver: fine, smooth, and combed back.

Before he arrived, my grandmother covered us with blankets and a recitation of the Sura of Ya Sin. When he asked about us, she answered him, "Let the little dears sleep."

4

The top was thrown in a ditch. My father trampled the mud houses. The beads were scattered from my waist and strewn in the ditches and corners. He trampled the rest underfoot. The girls stumbled confusedly as far as their houses. The boys took shelter behind the telegraph poles. Adil and I crouched between his arms, poisoned by the fury he exhaled from his pores. His voice rang out, reaching the farthest houses. He threw us in the middle of the house:

"The little bitch dances and sings in the street and the children hug her! I don't know what's going on behind my back!"

My mother stood in the doorway of her room, terrified. She coughed and pounded her chest, noiselessly. My grandmother and my father's sister came out of the room and stood in front of him. Like a sick bird, Adil clung to my mother's clothing, and I got up and stood up, between his kicks. I grabbed my father by his shiny boot and used it to crouch between his legs as he moved me around, grabbing me on one side and pushing me in the other. The floor of the house was my master. I was trapped by his voice, which came at me like bullets.

Whenever we saw him coming or leaving, he went from being an image of a father to being the Lord in his prime. We found the only way he relaxed completely was when someone was in front of him. It was always me. I provided an outlet for his talents, from his uniform to his lethal weapon, to his boots, which abolished all dreams: "No, Daddy, no, please God, just this once."

He did not frighten me the way he frightened Adil and my mother. At moments like these my brother went mute, not even breathing. He peed on himself, and when my father heard the sound of his peeing he roared with laughter. He left me for good, as if there was nothing wrong after all. He went to Adil, lifted him up high like a doll, and threw him up in the air and caught him, the drops of urine flying on to his hair and the tiles. My grandmother prayed and breathed on everyone.

When your father saw her, he changed; he calmed down. He loved and honoured her, and weakened in her presence. His sister, too, intimidated him. She went into her room, muttering, "If he knew how to raise children, he'd have raised himself first."

My mother was still standing there. I do not know who supplied my grandmother with all her authority, God alone, perhaps, or else she had assembled it all in her own special way. Adil was still flying up and down like one of his paper kites. My father's voice changed: "Look, the little devil is the only one who's not afraid. That's my little Adouli, his father's son!"

I was thrown on the floor, moaning but not crying. My hair was dishevelled, the ribbons falling out, my braids undone. I looked at my leg and rubbed it with my hand, and gazed at the squares of cheap tile. This one was a dirty blue; that one, a lustreless white. I calculated the number of tiles. I saw the anthills and the salty soil surrounding those little caverns. The floor surface was cold and damp. The shining boots stopped. Now Adil was in front of me and came to me and, burying his chest against me, he burst into tears. I tousled his hair and looked at his locks. I hugged him and he trembled, then broke into a new burst of crying. We cried together, giving it our whole voices, and my father pulled at me again.

"Be quiet. I'll get the belt and break your ribs."

He pulled Adil away and lifted him up, kissed him, and gave him five fils. He approached me, tugged at my hair, and lifted my head to face him. He took my hand and gave me five coins as well.

"Sweetheart, go and comb your hair."

Whenever his voice softened, the sound of my crying got louder. He pinched my cheek.

"God, if you don't be quiet – "

He kicked and slapped me. "This girl is a strange one. Does she want me to plead with her?"

Adil pulled me and got between us. My grandmother had not said a word. That was her; it was her way of pacifying him. My mother, in the back, took my father's attack in silence, a mythological creature stripped of all her roles.

Adil and I went into the bathroom. My father went into his room, his voice still ringing with every form of vituperation.

Adil shook my arm. "Huda, take this money as well, just be quiet."

I pushed him and he fell before me, got up quickly and stood in my face, pleading: "Huda, Daddy will be asleep soon, and we'll go to the blind woman, Umm Aziz, and we'll buy hot chickpeas and sweets."

I gasped and blew my nose. My grandmother was behind us. She stroked my hair and tilted my head to face her. I looked into her eyes, then buried my head in her concave stomach and hugged her round the waist. "Granny, what have I done? Why didn't Abu Firdous hit her for playing in the street? Why my daddy why?"

This grandmother was the centre of the circle. I do not know where she concealed her strength. When she walked, her feet, wooed with calm, did not touch the ground. When she spoke, her voice was clothed in caution and patience, and when she was silent everyone was bewildered by her unannounced plans. She was strong without showing signs of it, mighty without raising her voice, beautiful without finery. She was beautiful from her modest hem to her silver braids. She was slim, of medium height, a narrow black band round her head, whose ends dangled by her thin braids, which were white. I never saw anything as white as her complexion. It was a white between bubbly milk and thick cream. Her eyes were grey with dark blue, wild green, and pure honey-coloured rays.

When we saw her in the morning as we got ready for school, they were honey-coloured, and by the time we came home in the afternoon they were blue. But at night they were grey.

She was a well-organized woman; she loved justice and set great store by it. She rebuked my father and scolded him behind our backs, suddenly setting upon him, taking all her time, scattering him and tearing him apart, exposing him anew to us. She dazzled us every time she told us, in a clear, distant voice, as if coming from an abandoned dungeon, a story of my father, which she had never told before. She wiped the dust from the picture book and opened it. At the beginning was a picture of our venerable, terrifying, handsome, harsh, sceptical grandfather, who was in love with her, was jealous, and who never once in his life told her "I love you." He wore a tasselled fez and went to work in an office in Ali al-Gharbi, a village on the Tigris River. He walked around with a superior air, like an Ottoman pasha. When he went to work everyone scuttled out of his way.

Her fingers rebraided my hair. I was sitting on the carpet in our room. I turned my back to her, and she enclosed me between her skinny thighs: "Relax a little. You keep moving. Are you sitting on a fire or something?"

Adil was in front of us, holding the basin of water in which my grandmother was soaking the wide wooden comb. She began to comb my hair and talk:

"Your father fell on his head. His horse threw him while he was training – this was in Ali al-Gharbi – with your grandfather. The weather was fine, and it was a new horse.

"He used to take him out every day, before sunrise, and have him lead the horse and ride him. The first day he went, and then the second, and the third. For two weeks he trained and came back. He had changed, I do not know how, but he was different. His skin became tight, his voice had changed – he was like a beast of prey. His father sent him out at night and he wasn't afraid."

I interrupted: "Granny, you mean if a boy trains on a horse he becomes nice-looking?"

"Not just nice-looking! He becomes a man!"

"And if a girl trains on a horse will she become more beautiful?"

"No, a girl's beauty is in her silence and modesty. Do you want to ride horses as well?"

"Where are the horses now?"

"What happened then?" Adil interrupted.

I looked into his eyes. He was smiling, and I pushed him with my hand. The water basin spilled on his clothes and the floor. He did not get angry, but replied, "Stop asking so many questions."

"Fine, where were we?"

"My father was 'like a beast of prey,' " was Adil's prompt response.

She sighed a little and went on.

"He was not afraid. He was a little man. When he came back at night his eyes were still watching the sky. When the moon rose and the sun went down. He said the sky had many gates, all of which were open to him, and that only he could count them. He predicted so many strange things."

Adil interrupted: "What does 'predict' mean?"

"Imagine! You don't know?" I said. "Predicting means telling the future."

"Fine, my little Adouli, the sky was open, and he could read everything written there. He said your grandfather would die of drowning, and believe it or not – two years later the ship sank with six employees on board, in Basra. He said he would marry several times – he said that when I was running after him – I wanted to beat him. Oh, those days are gone. Only misery is left." Her voice changed and trailed off, to the Shatt al-Arab, and her first nights of watching over her son. She took the ribbon from Adil's hand and continued:

"He was fifteen, and the things he said frightened even me. I began to be afraid of him, but the third week they brought him carried on their shoulders. He was unconscious. He was sallow, stricken, like someone shocked by electricity, neither sleeping nor dead. There was a little wound on the top of his head – the skin was broken and the flesh had opened, but there was not a single drop of blood coming out of it. He was different from that day onward. He entered a new phase. Even he was scared of his imaginings. You know your father married before your mother; his first wife was with him a year and then died in childbirth, she and her son."

I asked her: "How did that happen? I don't understand you. You mean he went mad?"

She yanked my hair sharply.

"Oh, if only someone would cut off that gabby tongue of yours. No, he changed when his wife died, he changed completely. He used to stand around and lecture people, and curse the Regent and the English. "And he learned how to drink. At first he drank secretly – he was afraid I'd find out and get cross with him. When I found out, he began to drink in his room or in the bar nearby. At night, the local men brought him home to the house."

Adil moved off a little, leaning against the wall in front of us. My grandmother took the second ribbon and grasped my hair, and in a tone of voice I had not heard before, Adil asked: "Who gave him the pistol?"

"I asked him to enrol in the police academy. It only takes a few years, and they graduate you a police commissioner, then they promote you to assistant police director. He waited until the end of middle school. He was failing only the easy classes. Adouli, dear, everyone who goes to the police academy must have a pistol. Huda, sweetheart – "

She took my head and turned it toward her. She held my face in her palms and looked into my eyes.

"He is ill, and your mother is ill. We are all ill. You hear the way your mother coughs at night and spits blood. God forbid if – God forgive my tongue! – I'm not afraid of death, God created us and he takes us back. But there is still some patience. Your mother will travel to Syria for a little rest and breathe some good air. Your father's sister is still young. We are all waiting for Munir Effendi. Abu Munir died and left him the farms and shops, and he's starting to fritter away the money. He has no brothers or sisters. He is lazy and idle, and the girl cannot marry a stranger. You and Adil are the apple of my eye – you're the children of that dear sweet woman who has never said an unkind word. Poor thing, Iqbal!"

She hugged me, her arms tight around me. I kissed her and hugged her, burying my head beneath her ribs. I felt her belly, her soft breasts, and her long, narrow neck. I raised my face to her calm, sorrowful, inspired face, which never scolded when I was bad, but which was always responsive when I was sorry.

She tamed us one after the other, without our shedding a single drop of blood. She shared her thoughts with everyone, trained us without threats and took us to her bosom without menace. She prayed over us when we were ill, and fetched us from the end of the road if we ran away. She stood guard at the gates to our souls when we erred. She changed us with every passing hour. She did not interrogate or cross-examine us, or get defeated by our young naughtiness. She always said:

"If you do a good deed for someone, don't talk about it. No matter what happens here at home, tell people 'We don't know.' If someone tells you his secret, don't ever repeat it. A secret is like a treasure, and has to be hidden in a well."

And so on and on. When she went to the market, all the shop-owners opened up their secret rooms and new sacks of merchandise. They gave her the finest grains and the freshest vegetables, the whitest sugar, the purest rice, and shelled lentils. They put all her groceries in clean bags and sent them after her. She did not have to pay the price of all she bought, nor did they put her name on their list.

She paid on the first of every month. She was never late, and never haggled or procrastinated. She hated debt:

"God does not want any of us in debt to another. Debt shortens your life and blinds your eye." At her breast I mixed her good with my evil.

I gave voice to all my sorrows and dreams, and never feared any punishment from her.

I might disguise myself in other clothes, but to her my bones did not lie; my soul could not deceive, and my head would not bow.

"Dear Huda, she just kisses you, and has never once told us 'I love you.'"

"No one knows my Huda as I do. God keep you and keep evil far away from you. Now come iron my clothes – tomorrow I'm going to the General Retirement Directorate."

I did not know what this end of the month would bring. But my grandmother, my father's sister, and my father knew very well. My grandmother dressed up in her best clothes and combed her hair carefully. We brought her a large basin of hot water and the wide wooden comb that she pulled through her fine, flowing locks.

"Every day a hair falls out of my head. That's all because of sorrow." She switched her eyeglasses with the old black round frames for her gold-rimmed ones. We knew all these rituals from previous days. Everything was familiar; the new cloak came out of the bundle and was ironed, along with the only silk dress, with its design of graceful trees. It was ironed last. The high-heeled slippers were taken out of their box hidden in the bottom of the closet. That night my grandmother was transformed into a princess. Everyone was waiting for her blessings, her gifts and money. The General Retirement Directorate in the crowded Baghdad neighbourhood of Bab al-Mu'azzam was waiting for her.

When we set out for school in the morning, we knew that the retirement pension had been distributed. There was a chicken in golden gravy and red rice, fried aubergine, and plates of radishes, cucumbers, mint and lettuce placed all over the serving platter. There was a tall pitcher of fresh laban. The delicious smell of the cooking made me raise my voice. In school I told Firdous, "We have chicken and red rice at our house. You love it – come and eat with us."

We did not have one time for dinner and another for lunch. We ate when we were hungry. We knew that money was scarce. Our father gave a share, and our grandmother had to provide the rest. With this and that, we had curdled cream for breakfast every day. We had eggs once a week. My father's sister arranged all the vegetables on the platter, saying, "These vegetables purify the blood. Look at your face, how sallow it is."

We wanted more blood, whether pure or foul. It was not important, knowing that my mother's blood was infected.

I was not afraid of my grandmother's stories about her. Despite her absences and coughing, to me she still seemed young and strong, and a little older than my father's sister.

Whenever I asked my grandmother how old my mother was, she laughed and answered, "By God, I don't know. When she married your father she was in her twenties. She came from Aleppo with her brothers and her mother. Her father died when she was a girl. Her brother Shafiq was a doctor at the clinic in Karbala. He was quiet and gentle. Your grandmother did not let him enjoy life. She was strong, and had a sour disposition, God rest her soul. She always said, 'My son is a doctor and I must marry him off to a woman with money.' God rest his soul, he listened to her and worried that she'd get upset with him. Shafiq died a sudden death, before he turned forty."

"And my Uncle Sami?"

"The day we had the betrothal to your mother, he shouted and cursed. He said the girl's marriage was a shame, but Shafiq, God rest his soul, he said, 'Jamil is a nice boy from a good family.' Your grandmother died three years after he did. She suffered a lot – she thrashed about like a fish. She didn't die until God took her two months later. That left Sami, Widad, and Inam, and they stayed in the house as if they were his servants. He beat them and cursed. Your mother was the sweetest of all, like a rose. She spoke little. She was gentle and calm and never harmed an ant. Be merciful to her, dear God, most Merciful of all the merciful."

5

My mother followed my father to her room. They were face to face. The air in the room boiled with his shouts. She was standing, worn and weary; if she approached a sensitive point she would get burnt, and if she retreated she would be choked. His words came in a torrent, like a tumultuous wave: "You've all turned my hair grey. That daughter of yours is going to drive me mad. Everything is against me. I'm alone in Karbala. In the morning my boss shouts at me, and in the afternoon there are the screams of the prisoners. At night I do the screaming alone. Listen. I am going to get married. I have no more patience for this situation. I want more children. You finished by having Adouli. I want a real woman. I've given you my best years and my heart's blood, but all in vain. Go back to your family. Go back where you came from."

She said, between her tears, "Is this the truth, Jamouli? Are you really going to marry again? You are my family. Your mother is my mother, and you are the father of my children. How can you? Your children are living with their father's wife."

She knelt before him and trembled so that her teeth chattered. She sobbed. She reached for his legs and grasped his boots. She removed them and placed them side my side. "A woman may fall ill and take medicine and get well but she should never be left. Good God, Jamouli. Is this my reward?"

She began to massage his toes and leg in order to rise up. She removed his socks and smelled them. "You always smell clean. Darling, really, are you going to marry again, Jamouli? Do you swear by your father's soul?"

He pushed her against her chest, and she fell backward.

"Why do you want me to ask you for permission? You've been ill for years. All that medicine and all the expense, and you're still the same."

He stood up and began to undo his leather belt. He held his pistol and pulled out the cartridge clip, and placed it at a distance in the middle of the table.

My mother was afraid of every sort of weapon. She did not look at him, but he bent over her and raised her head to him. They looked at one another. His face was calm. At that moment my mother was able to get close to him, and before he removed his trousers he knocked her to the floor and threw himself on top of her. Her tears flowed wordlessly. He checked to make sure she was not dead. She knew he could not wait.

Amidst her tears and his murmurs, she sobbed, "Don't marry, Jamouli, please, God bless you, for the sake of the children, and your dear mother, who was better than my mother." He stifled his shout in her quiet breast, then stood up, preparing to leave.

"Now listen well, Iqbal. A few months ago I married a nurse from Karbala. She came with me to Baghdad, and she is pregnant. I don't like doing things illicitly. There are as many women as we have prisoners after me. They're young and pretty, and my boss had his fill of them. I swear to you, he even slept with the animals. Listen – don't shout and don't cry. You are going back to Syria, and I am going to stay at the prison by myself. You know the prison. Come there and see how it would drive you mad. Don't worry about the children. They will stay with my mother and my sister. Now get up and draw my bath."

"But Jamil, what about later on? What if I get well? Jamouli, what will you do later on?"

"God is good. You will leave here and come back safely. Now get moving – I want to wash and eat."

She burst into a fit of coughing such as we had never heard before. The sound of the wardrobe with the three doors erupted in its own fit of creaky coughing also. When we opened its warped doors we could not shut them again, unless someone pushed them up.

My father left it open, having taken out his large white towels and gone out.

This room was at the end of the hallway, far from us. It was the cleanest and warmest room, its walls painted a light blue. An iron bed stood in the middle, and the wardrobe took up most of the middle wall.

Also in the middle stood a mirror which had lost its quicksilver backing, and its wooden frame was worn at the edges. In one corner was an old chair and earth-coloured table where my father's shaving things were set out with a bottle of aftershave and one of rosewater. A Qu'ran rested on a small shelf covered with a cloth embroidered in white thread. In the other corner stood old shelves upon which books were arranged: Dar al-Hilal editions, the Reader's Digest in Arabic, and the stories of Jurji Zaidan, Taha Hussein, Tawfiq al-Hakim, al-Manfaluti, and issues of Egyptian magazines such as *al-Musawwar*, *Akher Sa'a* and *al-Kawakib*. The only window, which looked out on the courtyard, was usually closed. When my father was in Karbala, its yellow curtains were open. The glass panes were always clean. In the summer, my mother wiped them with old newspapers, and in the winter she wiped the traces of rain away with a dry cloth. The floor was covered with a long old carpet folded in more than one place to make it fit the small room.

My mother wandered about, giving off a scent of defeat. She stood in the embrace of that heritage. The boot, the pistol, the madness of this rupture. Her first indifference came to an end. These changes had taken place behind her back. It was not important now that she change her name or blood type; nothing could bring back the past, the magic or her beauty or her serenity.

She paced the room, and I paced with her behind the window. She was agitated, facing all the objects and things, looking at everything around her as if seeing them for the first time. She walked unhurriedly, touching the Qu'ran, fondling it with her hand and saying, "They left me in your care. You beat me and cursed me."

She staggered, looked at the carpet and the open wardrobe. She fingered the bookshelves and her muscles contracted. She snatched the books and threw them to the ground, shivered and sweated; her face grew paler and the familiar objects became masses of hidden meanings. She knocked them to the left and the right, and stood in front of the mirror, advanced slowly and opened her mouth in an obscene movement, lifted her hair up and then let it fall on her face, moved her arms. Her eyes bulged, as if she were emptying her bowels. She let out a cry and put her hand over her mouth, slapped her face and tore at her hair, and caught her breath sharply before the mirror: "Is it true what Jamouli said? My face looks frightful. My God, I'm afraid of seeing it in the mirror. I've been afraid of that face for so long. I certainly was the most beautiful girl. Oh. God forgive me. Where is my mind? Mama, come and look at Iqbal now. Jamouli is married, Mama. God Almighty. He married her and she is pregnant as well."

She smacked the surface of the mirror and dropped to the floor. She opened her legs and beat on them. She raised her nightgown from her slender thighs and scratched them. I could only see her undulating movement as she shook and hugged herself, as she raised and bowed her head and back before me.

"What do I have left? I will never see the children again. Mama, come look at me now. No, no, let me come to you instead. I would love to travel there. I will see you and my brother Shafiq. I will be able to tell truth from falsehood. Poor Iqbal, humans get ill and are stricken and rise again. They must pick themselves up and stand tall. Death is an attitude. Why, Jamouli, why? Is she better than I am? I am the mother of your children, the mother of precious Adouli. Oh, Mama, who will wash Adouli's head and prepare sweets for him? Where will I go now? Jamouli is trying to drive me mad before I die, and I swear to God his dear mother is the only reason I have stayed."

I heard my father's loud voice:

"Iqbal, come rub my back."

She jumped up suddenly as if stung. Her voice was inaudible, smothered by tears. She opened all the doors of the wardrobe and started there. She took out my father's clothes, his new uniforms, his ironed shirts, his hanging ties. She threw one uniform after another onto the floor, scattered the shirts, and hurled down the ties, like a genie the hot earth had produced, or who had flown out of an oven.

She shrieked and crawled. She snatched the clothes and threw them away from her. She turned and curled up on the ground, then stood up. She turned about, flushed with anger. These were the clothes of the long nights of waiting.

These were the shirts of the only man who had ever known her pure embrace and sunk his beak down to the ailing roots. She had pulled him away with her hand, rubbed his back, chest, and hips, his thighs, legs, and feet. She had seized him by the arms and gone up to his head.

She had whetted his appetite for sleep and snoring. She had covered him and gazed at him. She had sat at the end of the bed until he awoke, and when he called to her she went to him, bruised but radiant.

This was the bed where she had learned he was a man, that he was the ruler, the father, and the chosen one. She trod and leapt and wailed. She pulled out the white undershirts and held them to her face, smelling and kissing them. She held his underpants, his white and blue handkerchiefs, and his socks, and moaned, "Jamouli is married and he's got her pregnant! Oh, no! What shall I do now?" This was the first time I heard my mother's voice torn out of her like a rope lowered to all of us. It cut through the walls and our ears. It was nothing like our voices or our daily quarrels.

The voice started and awakened, stopped and then rolled on, carrying a banner high, stopping before me in the window.

She did not see me but I saw her. She screamed in my face: "Go! Get me the scissors."

This was my father's precious inheritance, his bed, his clothing, his bloody receptacles, his insides, his madness, the emblems of his police work, the conditions of his good looks and elegance.

My father spent most of his wages on clothes. In winter he wore grey and black, and in the summer blue and beige. He glittered and shone as he stood in front of the mirror treating his glossy hair with a special white hair cream. He covered his face with cologne. On his body his clothes became like wings – he seemed to fly out to the street, and his mouth watered when he saw himself in the eyes of the neighbourhood women. He shivered as he placed his watch with Roman numerals – it had been a gift from his grandfather in the days of the British – on its gold chain and hung it from his waistcoat, letting the chain gleam and flash across his stomach.

He left the house alone, walking like a king – he had trained himself in this walk. He never bumped into anyone or greeted anyone with any hand movement. His fingers were in his pocket, his leather belt, the opening of his collar, his crisp trousers. He never took a step out of his way. He never lost a button or dropped a handkerchief. When he boarded a bus, he rarely took a seat, though when he did he made a great show of positioning his arms and legs. He held his breath, his arms folded tightly against his ribs, his skeleton perfectly erect. He took his uniforms to Abu Ghanim's ironing shop himself; Abu Ghanim ironed the clothes of the rich families at the far end of the neighbourhood. He picked them up himself too, felt, sorted, folded them, and hung them up in the wardrobe himself. He ordered my mother, "Wash them separately, and spread them to dry in the shade so they won't fade." We never dared touch them.

He called out, "Where are you, Iqbal? Come wash my back."

I did not move; my head was splitting with her screams and sobs.

She paced around the room, bent over, straightening up and taking whatever was in front of her, tearing it with her teeth and throwing it on the floor. She hiccupped: "I won't die twice, and if I die now, I'll die contented."

I wept in the courtyard. My grandmother, aunt, and Adil were walking in front of me. They went in to her, and now her voice was louder than my father's. "Your son is married, Mama, Umm Jamil! Jamouli is married and the woman is pregnant! That's your reward. But now he'll see who Iqbal is."

His voice, our voices, her voice – all had drunk from the same river of madness and grown in the same house of utter ruin. Adil was squatting in a corner, watching and crying. My aunt picked up the clothes and books; now the room was starting to resemble the messy room high up on the flat roof.

Farida wailed: "God protect us from this day! God will kill you,

Jamil. Huda, come help me clean up before he comes in here or blood will flow tonight."

Alone, I watched, and watched, and watched, and stumbled and bent over. My grandmother cradled my mother and hugged her tightly, prayed over her, and pulled her by the arm. "God is great, my daughter Iqbal, God protect you, God bless you, now let's go, let's get out of here before – "

My mother screamed at the top of her lungs. "What will happen now? He'll kill me for tearing his clothes. I don't care! I'm dead! I don't even have any blood left! Jamil is married. Huda, your father is married! Come, Adouli, you have brothers! Adouli and Huda, come and see, today we will have this out. I want him to come here, in front of me. Come out of the bath, Jamouli, come here and see how Iqbal isn't afraid any more. Mama, everything is gone now."

She coughed, and for the first time I saw her blood. I cried out, and so did Adil and my grandmother, and for the first time I saw my grandmother's tears.

My father's voice: "What's going on? Iqbal, Huda, Adouli, Farida – where are you?"

His voice approached and my grandmother dragged herself and my mother by her arms. My mother resisted and tried to squirm away, ablaze with rage. Her voice split the air: "I won't leave! I want to stay here and see him. I want to die today. Adouli, Huda, come near me. What more can happen to me? Can it be worse? Where do you want to take me? This is my house and this is where I'll die!"

My grandmother put her cloak over her head and over my mother's as well, and pushed and pulled her, breathing prayers and murmuring "God is great." My grandmother forgot nothing. She took her suitcase, put on her spectacles, covered her face with her veil, and pushed my mother in front of her into the street.

My father was now out of the bath and stood wet and frightened. My mother's coughing sounded from beyond the doorway and was heard no more.

Cough now for as long as you like. We will huddle together, Adil and I, and cry, and gasp, and say no more.

6

‿◉‿

Your father camped out in his room. He opened the window: "Farida, dear, please bring me tea with your sweet hand."

He began with the Qu'ran. The tea tray, bread, white cheese, and mint leaves. Your father's sister waited for these roles. She spun wool winter and summer. Your grandmother was on her bed, a Qu'ran in her hand. You children dared not play.

I did not like school, but even so I passed miserably at the end of the year.

Here I could not bear the silence. There was no coughing, no infected blood, no healthy blood. Even the ants lost their way in this house. No one quarrelled with anyone; no one stammered with anyone else. Adil opened his reading book, and dusk fell as he murdered the letters *ba*, *dal*, and *dhad*. You wandered about among them all. You undid your braids and toyed with them. You struck Adil and knocked the book out of his hand, and you trampled the paste for his paper kites. You wanted to hear screaming or a gunshot; no one had screamed or coughed for months. They said she was getting worse with every passing day; the air of Syria was doing her no good; she wanted to breathe the same air as this man.

Through the little glass window at the top of the house, the sky looked grey and black. Time was bewildering: it did not pass quickly, either to let you grow up or to consummate your despair.

You walked in front of the door to the roof which admitted an icy breeze in winter, and dust as fine as powder in summer. You used to stop up the cracks in it with thick plugs of wool.

You went up the stairs in the blink of an eye, and were in front of the door, and when you opened it the whole house shook. If you had remained standing where you were, you would have gone mad. You held the plugs of wool to stuff between the wall and the lock, and looked like a professional thief. The door had to open once.

You stood on the roof; there were no big clouds here, just the sky, and any deal you made with God had not been kept. You asked Him if you might share that traveller's coughing and illness, but He agreed only to multiply the quarrels between you and those around you. Your father was number one in this respect, and the sky crowded you and pushed you into war. There was no door before you. Where had your grandmother aimed her prayers?

In those years, your father alone was engrossed in planning for this family, producing his own public and private evil.

He married one-eyed Nuriya, the nurse at the government hospital in Karbala, and moved into her old house, to fight with the screams of her mother and brothers into madness. His visits grew more frequent. Every Thursday he visited us, and sometimes he came in the middle of the week. Debts became drafts drawn on the future as he waited for his seed to grow in the belly of his new wife.

"You have not seen her and you never will." That was your grandmother's vow. He had organized his own world, which began with drinking and ended with drunkenness.

They said that she had lowered herself with some of the men of the holy city and submitted to various influential characters. They said your father had fallen under her lethal spell. They talked and spread rumours, and your grandmother did not advance or retreat in her decision: "Listen, Abu Adil, as long as I have breath in my body, Nuriya will not set foot in this house. She is your wife – fine. The past is hers and the present is yours, and what comes after is your own business." His voice rose in grief and sorrow: "Mama! Are you telling me to divorce her? She's not a nobody. Her mother used to read prayers for Imam Hussein. Her morals, God forgive me, are like anyone else's, but she's a nice girl. I've lived with her, and she loves me and is very afraid of me. In a few months she'll have the baby. For the child's sake let her come here and kiss your hands. Please, Mama, God bless you."

She did not reply, or turn around, she only looked down. He went out with his head bowed. He set his table: peeled cucumbers, boiled beans, hummus with sumac and lemon squeezed over it. Three empty glasses. He always put out this number of glasses with a bottle of arak. He looked at them; he liked them empty but wanted them full. He whispered to the arak and joked with it; it waited for him and he waited for it.

Across the table, the father waited for lines of caresses. The policeman's despotism relaxed, and his official clothes came off. He traded his boots for the bare floor, and his bare toes trod upon it. Here he encountered disorder. He was out of prison, and did not harbour anything but love. He acted lovingly toward you each in turn, starting with Adil, calling him and joking with him. He hugged and kissed him, lifted him up in the air and buried him in his chest, then put him down. He put him on his lap, and they looked at one another. He started to read him a book, spelled out the words, and helped him with the arithmetic. He pinched his cheek, saying, "I can never get enough of you."

Abu Adil leaned his chair against the wall, spread his legs, drank and drank, nibbled one end of a cucumber. He drooped sleeping on Adil's plump legs.

"God bless you, Adouli, rub my head. My head always aches when I come here and when I go to Karbala. I feel as if there is a voice calling me. Every day I hear the voice, and every day the voice changes. It sounds like a voice I've heard before. I know it from afar and it frightens me. Adouli, everything tires me out – even sleeping makes me tired. Ah, that's where it aches, there, behind my ear. Dear God, you know when I hit your sister, I cry later in the train. In prison I remember your tears and her tears when I hear the prisoners screaming and crying. You know, Adouli, sometimes I think you should come to see me, there, in the prison, so you can see how I live. Dirt and black death, flies and lice. Locusts and rats are my only friends there.

"Ah. Every time I want to drink until I'm drunk, but every time wake up more sober than before. Your grandmother says arak is a sin. Yes, there's a lot of sin in this world, but if she tasted arak just once she'd get used to it like me. Don't be afraid of me, Adouli. I don't frighten anyone. I'm always afraid, but I don't want you to be afraid of anyone. Even God Almighty himself doesn't want just our fear. Adouli? Is it true, that I'm not frightening? Tell the truth. Don't be afraid."

He got up and leaned against the wall. Adil was silent, rubbing his fingers together and then raising them to his mouth. He chewed his nails and swallowed them. "Have you had supper?" He nodded yes.

"Go finish your homework. Come here and let me kiss you."

The call to evening prayer dispersed the voices, and you were consumed by weeping. You cried alone, and your tears made you laugh. The stars were unruly, and this whole horizon was a lie.

The floor was stained, warped, and uneven. When it rained, the rainwater seeped into the cracks, holes, and hollows in the roof of your room. You put out the buckets and heard the water plopping down.

This frayed laundry rope, that scattered and chaotic room, dusty and deserted, the door scorched, and everything in it old: pillows, blankets, broken chairs, boxes broken apart, copper and silver utensils, spoons and dishes. This was your grandmother's first dowry. She was in love with anything old; every year she came up here, spread out the contents, and began to clean, rub, and polish them. My mother was with her. We all came up here to see our grandmother's secrets; everyone in the family had a share of this heritage.

Open the boxes and look. Objects that have never been insulted, never been whipped with a lash. They are united in their dust, sleeping where they lie. They are rusty and faded, yet they cling to their silence and passion. They began to address me, to talk to me, and I asked them to confess. They are more beautiful than the others: my father, his sister, Rasmiya's husband, and Uncle Munir.

Things had this tremendous quality, of becoming pleasure; sleeping between the palms of my hands. My grandmother's silver spoon, the one she ate from on the day of her first wedding.

Your father would impose his tyranny on you if he knew you were up here. Your aunt, his sister, would hit you, your grandmother would be silent; your mother would not come.

Search and search well, and restore safety to all these things. Organize the converging paths and clear the way for seeking the pardon of all that remains before you.

You were here, and the only window, with its dusty glass, was before you. The neighbours' clothes were strung along the clothes-lines on their roof, cheap and dragged down by their wetness, touching the ground. The clothes were like people being hanged, and I was waiting for my father to guillotine me.

My father was the same size as me. Our fear of one another had no mask. He could not bear the loss of me, and it was the same with me. We attacked each other's walls, and did not confuse anything that passed between us. We plotted together, and publicly: the arena, that place of rancour and celebration, all this sameness. We spread out there and waited for one another.

They said, "Huda was suckled by Satan."

My mother had nursed me only a few days. I drained her milk; I drank only the *khishkhash*. There you beat longer and harder. They stood, one of them accompanying me to that tent: my father, and I felt as secure as a highwayman.

Night raised up its new inflection. This roof trained me to count the moths that entered my dreams. They entered the bodies and ate away at everything, as I remove one after the other: first of all my father.

The pistol threatened everyone. He carried it and went up behind me. When his fear exploded, we went limp with fear.

He did not pull the trigger. We encircled his footprint and went up to his waist. He was not heavy but he was tall, his shoulders waited for me and his face changed, he changed, smoothing all the paths for me so I might move toward him. Perspiration gathered between his fine, delicate nose and his pendent lips behind which his saliva was gathering. He sprayed it in my face and spat it out in the air between us, as Umm Suturi did in the baths. Then we touched, and at that moment hugged one another, and I pressed my face against his stomach. I clung to him with both arms, though I could not reach all the way around him. This time he was the one who kicked.

I surrounded him, I held him, I clung tightly to him and turned my face up to his and looked above the first blow and he was carried away to me.

He knew my braids perfectly. My hair ribbons did not defend me. The neighbours came up to the roof, growling. Mahmoud was silently weeping; Adil saw my grandmother not uttering a word, approaching, not resisting, but ready. If he overstepped, she would unleash her voice and her hand. The pistol was in his hand, and he was tapping it on your head. You did not cry. Your eyelids shone, your eyes were clear, and your eyelashes were dry. Curses were aimed at your back, and your head was lifted to the sky – the Baghdad sky seemed to belong to a bygone age. The world was like a round table on which your body was sprawled. Father started with the shoulders and descended to the restive arms, to the belly and buttocks. He brandished his pistol: "I swear to God, if you come here again, I'll kill you!"

At ten you confronted the first policeman in your life, your father. You summoned up all the sins of ten, the rashness and recklessness, the lies and tempting dreams, the yearning to get sunburned in order to shine more: get all this out of your ribcage and celebrate like the feast of Muharram. There I celebrated with the police and summoned to me the insects, black and red ants, and unknown things. The cavities of the locked boxes, I cut the strings of every fact in two, to see, and see, and see. There I opened up to him a fountain of the spirit and did not consent to kill him. If I killed him, who would straighten out my skull? If he died, who would I fight? If he went mad, who would quarrel with me?

Alone, he followed me to learn that I had surpassed him.

My father.

I turned and turned, and six legs stood observing me, eyes bulging out of their sockets without meaning, without hope, without grace, neither mourning nor laughing, nor shouting.

Under that sky my father took me to the gate of Hell; the future was a flaming ball exhaling hostility, its pores covered with blood, dirt, and fear. His voice soared, frightening enough to remove the hair dye from the neighbours' heads.

"You whore! What are you doing on the roof at night? Making

dates with the neighbourhood boys? Shitty Mahmoud? Suturi the pigeon boy? Cross-eyed Hashim? Speak!"

Speak, Huda, don't delay. Defile him, hunt him with your wickedness – you have no prey bigger than he.

Between the stairs you used to threaten him. None of them knew him as you do. He was the first inspiration in your life. Open your eyes and look at him well. Hold his breath, and share with him nothing but plans for murder.

For what was the celebration of the scuffle except to make your claws scratch more, your teeth bite more, your muscles attack more?

Steal the food which was hidden for him, sweets and fruit. Damage his books and magazines, read them and scatter their thoughts on him first. Pour out on him this glory from your strong little heart. Go to your mother on your bended knee, open the gates for her and seat her as the queen of death and life; weep for her, for she is dying.

I dried my face, fixed my hair with my hands and pulled it back, looked at my appearance and watched Mahmoud at the opening of the street. You were in the street again, and the children brought me back into their authority.

"Listen, I'm a boy as well. No, I'm not a boy, but I can be like a boy."

"But I want you to keep on being a girl," replied Mahmoud.

You hated this admission of his, but loved it too. It was clear from the beginning – you were always this way. But I loved rebellion and the friendship of boys.

I knew that if Mahmoud and I pooled our strength we could utterly convulse this neighbourhood of ours.

"My mother says you're like the devil."

"Listen, you give me a headache with everything your mother says."

I laughed, and he looked directly at me: "You're prettier when you laugh."

I look at him, still laughing: "I don't know anything about the devil, but listen. You're with me, so that means you're with the devil. Agreed?"

7

$\sim\!\circled{6}\!\sim$

"In the name of God, the Compassionate, the Merciful, open the way."

"Dear Abu Hashim, put the bag down there in the courtyard. God give you rest. Yes. here."

He raised his head a little but did not look round. He swung the large bag down from his back. My aunt stood at the window of our room. Adil and I were in the middle of the courtyard.

"Now smoke a good cigarette. This is first-rate Abu Nuri tobacco."

"Listen, as soon as I finish rolling them, I'll send you a dozen that will do you all summer. You deserve it, Abu Hashim."

"God bless you in this life and in the hereafter. All the shopowners say that Umm Jamil is a fine woman, a religious woman without fanaticism. If she prays over a wound, it heals. May God make more like you, Umm Jamil."

He said that and walked by the door to the courtyard. Uncle Munir stood behind the threshold.

"Welcome, Uncle Munir."

He did not look or respond. He entered. Adil disappeared from his path for a few moments, my grandmother turned her back to him, and I stared at him. He knew his way. He went into my aunt's room and stood in the middle of it: "Always at the window – don't you get bored? Every day the same lampposts and the same view. I'm here in front of you now, and you're waiting for me. Everything will be fine."

She did not turn or respond. She moved from in front of him, and before she passed him he pulled her to him: "Where are you going? Are you upset with me or just being spoiled?"

"Huda, my girl, bring me the chair and a tray, and you and Adouli come and help me a little."

She pushed him away.

This was the tobacco season. The pure tobacco was the colour of the smoked sun, sifted and milled. Each season my grandmother prepared her cigarette-making tools. We brought her the low chair and the round copper tray, and made her a place to sit, with cushions behind her back and a blanket over her lap that reached to cover her feet. She had bags of paper to her left and a sack of tobacco to her right. When she scooped the tobacco, the thin, fresh white paper was ready, cut into shape to roll into cigarettes. The fragrance filled the house, and the tobacco dust filled our throats. She had a fit of coughing, but her fingers worked, folding, rolling, and twisting each cigarette, then she licked them sealed and counted. Every so often she removed her spectacles, polished them with the hem of her nightgown, then replaced them on her nose.

"If God would only have mercy on me and let me stop smoking. My chest is swollen and my breath is short, but I love cigarettes, God curse them and the day I first tried one!"

"What do they taste like, Grandma?" I asked.

Uncle Munir let out a bark and said, "What is it with these homemade cigarettes? I tell Umm Jamil, why don't you try Craven A?"

He removed one shoe and turned to us; he was sitting beside our aunt on the carpet.

"I don't like the English or their cigarettes. God damn them in this life and the hereafter," she said.

He let out a loud laugh and began to ruffle Adil's hair. Abruptly Adil ran away from him. "Now time is with the English. You've started to talk about politics – aren't you afraid? That isn't your way." She coughed and tapped on the tray in her hands to gather the tobacco together and measure it. She looked up at Adil, pulled him close to her, and said:

"Good. Five hundred cigarettes. Every time my dears come, they help me. Today they left me alone." Munir lit a cigarette and blew the smoke out on Adil's back.

"Why don't you answer, Umm Jamil?"

Our aunt was silent. Adil walked past Grandmother, measured with her, and threw the thick strings tight over every bunch. I sat in front of them all and watched my uncle. Today he was wearing a dark blue suit and a tie as yellow as ground cumin. His shirt was clean and his socks were black. He had a mysterious smell, and his laughter was thick. It was afternoon. The sun shone behind the glass panes like a ghost.

Our aunt was random in her anger. He was slow in proposing marriage to her, and held off, and manœuvred, and advanced, and she, sitting or standing, sleeping or awake, was waiting for a gesture of his hand or movement of his tongue: "I say, Umm Jamil, when will my cousin come?"

Grandmother did not look at him. She was that way; she knew from the vibrations, from the skeleton of each person who came in or went out, who was in front of her, whether he raised his voice or lowered it. She besieged him with her silence and apathy, and he stumbled as she had planned he would.

She knew this Munir well, when he mocked or stalled, for all questions and answers were clear: "Do you have anything new to say? Say it."

Adil stumbled after counting one hundred cigarettes, and his voice sounded high. He sneezed, blowing snot. "Grandma, haven't we done enough? I'm going to choke from the tobacco smell."

Grandmother smiled and raised her head to look at him. After your father, this was her king: young, cute, sneezing from the tobacco, sleeping when he was ill, falling silent when strangers arrived – he did not like this Munir following him up to the roof or in the street. When he slept, and when he awoke.

"Isn't there any tea? God give her peace and bring her back safely, Umm Adouli. As soon as I'd come in she'd set the tea tray in front of me. I say – what is her news now?" Adil and I are hurt. For a long time we have not dared ask about her. What was whispered was still whispered, and what was humiliating was still sealed in the rooms and our mouths. The day our mother's sister Widad came, they made us go outdoors. She kissed us and gave us each a squeeze, breathed in our faces, and looked long into our eyes without crying or speaking. She did not talk, nor did we ask questions. Everything pulled itself away from us, the absent woman, and travelling woman, the ailing woman, and the birth of our new brother, whose father named him Saad: happiness.

He came one day and felt assured that when the people saw him they would know that he had sired another son. He walked into the house and knew that everyone would maintain silence, our silence in the face of the newborn baby, and the long absence of Iqbal. Given all this, there was no one to mediate between Grandmother and Mr Jamil. Until the decision by the prison warden to promote him, saying: "This new star on your shoulder is a gift from Saad."

That sparkling yellow star brought us from threat to fulfilment.

Grandmother did not take a step toward him, and he did not move away from us. She did not deny his marriage and she did not give it her blessing. She did not reject his new fatherhood or oppose it. Now he took long absences from Baghdad and from our house in al-A'dhamiyya. He sent his monthly salary by courier. His wife sent sacks of luscious Karbala dates in summer and baskets of oranges in winter, chickens and cheese, loaves of bread fresh from the oven, and we all ate it – except Grandmother.

Our aunt arose lazily and went into the kitchen. We heard the sound of water, the glasses, the low murmuring as Uncle Munir got up behind her: "Munir, wait a little, Farida's coming back. So what have you decided?" He sat and grumbled, bumping into the tray and knocking down the bundles of cigarettes. Looking him right in the eye, I heard my voice say, "Oh, fine! Your eyes are open but you can't see. Good God!" He laughed and restrained himself. "Today I'll let that pass, but I'll cut out that nasty tongue of yours, not now – when I come to live here."

"You know, Uncle Munir, when you get cross you make me laugh. I swear."

Adil and I laughed.

"Grandma, why does Uncle Munir want to come and live with us?"

"We'll see when he comes and lives with us. My father never agreed to any stranger living with us."

"Grandma, an uncle is not a stranger."

"Living with us" – these are new words which even Grandmother was using.

"Listen, Umm Jamil, I'm going to build a new room on the roof, and renovate both rooms, and paint them. Next week the workers will come. We'll bless the rooms with a reading from the Qu'ran next Thursday. I'm tired of being lonely and alone. My house will be too big for us. I'll live with you. I can keep an eye on the children if Abu Adouli is away. I'll live on the prayers you say for us all. Is this a bad idea or a good one? If you have different ideas, tell me." For the first time she lifted her head and looked at him. His face displayed every contradiction. This was Munir; he took my grandmother by the hand and led her up to the high roof. She stood there, her head tilted up to the sky, shy and radiant, as beautiful as a fairy. Her dream was before her and relief was drawing near; her joy, though, was postponed for one month.

He kept her there and answered all her questions. She wanted Farida to have virtuous children and constant protection. She wanted to stay with her here, to keep her company and await her first birth. Farida was the last of her children, a lover of new ways and fugitive dreams. Farida and Munir: my grandmother laughed, she laughed loudly, and every inch of her seemed delighted.

"By God, if I knew how to trill, I'd wear out my voice for good, but even if I did know how, my tongue would be bloody for my darling, Iqbal. Abu Adil has no objection; you know him, he's short-tempered, but he has a good heart. He only loves virtue, and God blesses all good things. As far as what you said, I'll do as you say, don't worry, and congratulations, dear Munir."

He rose from his place, took her hand and kissed it, and she kissed his head.

"Ugh, you smell of arak. You drink in the afternoon. God forbid such a thing."

"Umm Jamil, that isn't arak, that's the spirit of life."

"Quiet. You are like Abu Adil, you always have an excuse ready. God guide you and heal you!"

Farida stood before us, the tray of tea in her hands, a tumult of joy on her face. Everything was clear. At last it had been said openly; the female had craved this glory.

Farida smiled with a shyness that did not suit her.

So let males marry females; let your aunt launch her fireworks into the vast sky, and let it show anew in your faces, and the faces of the neighbours, the coffee-house men, the women at the baths, in the faces of the neighbourhood youths and in the street. Slither down on the high wedding throne, children, and read her the thousand commandments. Let Farida, electrified by her constant laziness and long mistakes, walk to Uncle Munir. Let her rock back and forth to the music if her head is bowed or her hand is bound. Let her swallow his saliva, his water and his phlegm; let the first Farida disappear.

Rejoice, now, and strew flowers about her and about us. Sit on the threshold as Rasmiya does, as my mother did and my grandmother does, and wait with the long queues for his bald head and vomit. Go to him. Let Rachel create a wedding outfit for you.

"Take me with you!" Let them take that Huda, and to carry a sack of clothes for you of brilliant colours, white, pink, and violet.

Your grandmother said, "Dear Farida, make a violet dress – you know how much I love that colour."

We went to the markets in Baghdad, hand in hand, face to face with the city I did not know. We rode the British-made, double-decker red bus. We went to the upper deck. Just look at this limp, effeminate Tigris! I loved only the Euphrates.

The first time I saw the Tigris, I ran toward it. Its torrent ran deep and the mud was thick. The water was cold. Grandmother took us to Ali al-Gharbi and left us there. We visited the old house, and saw the women and children, the girls and boys, the donkeys and chickens. We walked over the cattle dung, picked berries from the tree whose branches drooped to the ground. The faces around us were sweaty and dark skinned. The boys wore short, torn dishdashas, and the girls stared at me as I put shoes on my feet. They touched the ribbons in my hair and laughed, winked at one another, and hung on me. They felt my dress and my hand. I was in their midst, glowing from the crown of my head down to my toes. Grandmother left us and went into one of the houses. I walked among them, and Adil stood as if pinned to the ground, afraid of new places and people. I did not bother with him, but left him standing there and went to play with the boys. I took off my shoes and loosened my hair, and we all held hands and ran to the shore. They drank and I drank, and we splashed water on our clothes and in our faces. We laughed and shouted and waded in. I walked, headed away from them. The water seized me entirely. I spread my arms and embraced it, immersed myself and got knocked over. The boys grabbed at my wet hair and pulled it, shouting. I kept going, still wading, as if waiting for someone to emerge and talk to me.

White birds pursued me, their wings beating in the leaden sky. These were birds I had never seen over the Tigris. They were beautiful and shining; their legs were red and delicate, and their feathers were clean. They moved like dancers. They glided and collided with the water, and the sound of their beaks tearing at the small fish took me to Mahmoud speaking to me, coming down beside me, flapping their wings, slowing down and drinking the water, and looking at me. Everything took me into its embrace; the embrace of the water, of the birds, the touch of those excited hands in the middle of the Tigris. I shouted to Mahmoud.

Firdous said he was ill, and she surprised everyone and went into his darkened room. His mother was at the market, and Firdous was beside me. I was standing at his head, and for the first time I laid my hands on his flesh; he was burning up, and so was I.

Mahmoud, you have suddenly grown two years and waited for me. Leave the fever behind and come with me. Mahmoud, my mother's lungs are diseased, and I am consumptive because of you and me.

I took his hand and folded it, smelled it, and kissed it. These were the fingers of the first man in my life.

Mahmoud did well in school and I failed the exam. What was it about the exam and the school? Answer me, Mahmoud.

I wiped away his sweat and looked at his beautiful face and his fine, curly blond hair. His cheeks were fiery hot and briny, his lips were dry. My tears did not fall. You both agreed that neither of you would cry, and you wrote it with black coal on the public street, on the roofs of houses, on the carvings of the houses. No matter if you both failed in school, or if your fathers died or your mothers went mad, or everybody committed suicide, or if our brothers were killed – we would not cry.

You looked at his arms and entwined your arms around his, and asked him to laugh and be bad. Mahmoud laughed, and Aunt Farida will get married in a few days; Uncle Munir will remove the warts and lance the boils. And my mother was still dying.

I leant over and kissed him. Firdous cried silently:

"Huda, I'm afraid he'll die. Typhoid is killing people these days. All that's from swimming in the afternoon. Tell your grandmother to pray for him until the fever goes away."

"He won't die." We had agreed that neither of us would die before the other. We had not actually said that – we did not know how it could be said.

That Euphrates came out of its haughtiness and tied me to its horizon; and this Tigris, behind which I saw no horizon.

Men crossed the old wooden bridge, and women wrapped in black cloaks; girls in school uniforms, children marching behind their muttering mothers. And the bus that is capable of taking you to the end of the earth.

Before going out, my grandmother said, "Buy the household things first, and later on the wedding clothes and gold."

Mahmoud's mother's voice was behind me, and his voice was before me: "Huda, Huda."

My aunt pulled me from the high window: "Come on, we're going to the market."

Every time, one of my mother's or father's sisters came with us. Today it was Aunt Naima's turn, the friend of my grandfather's big house, the companion of her sister Aunt Bahija, and the seamstress for the homes in the other streets.

She was tall and intimidating, and her eyes were as black as charcoal; they were narrow, and their whites gleamed, and their irises were the colour of roasted coffee. Her nose was straight, her lips thick, and her hair as curly as an African's. She had a strong body and moved quickly, and her voice was calm and tender. We loved her when she took us and began to tell us stories and tales, but we could not stand her when she was angry. She changed all of a sudden and went into fits, she shook and trembled, her eyes grew wide and her hair fell loose, her fingers became stiff, and she rent her clothing. We looked at her stretched out on a sofa in our grandfather's big house; everyone but Bahija Khan had vanished before her. She stood at her head, dabbing her with cold water, fanning her diligently. Then she bent over her, took the fingers of her hand and began to massage them. She looked at her as if seeing a creature that had descended from heaven to be her guest alone.

The two sisters stayed that way until everything disappeared, hugging each other in silence. Aunt Naima had never married, and now she was past forty. From the refuge she took a girl to wean, named Zuhur. She was modest, obedient, and tender; she sewed her the costliest clothes and waited for a bridegroom to come for her.

✿

This was the day for scouring the markets of Baghdad. I changed my rhythm, leapt and played, ran away from their hands and stood alone for a bit on the Old Bridge. The buses passed, and I stood listening to the voice of the corpses colliding: the English, Nuri al-Said, the demonstrations, the firing of bullets, bodies lying on the bridge while others fled into the river. Mr Ghanim, the son of the barber in our neighbourhood, was brought here, carried on their shoulders. Bullets had hit him in the back, gone down to his pelvis, and not come out. His right leg was paralysed. He remained sitting in his father's shop, behind the table, collecting money, writing down names, and cursing the English. He had left the school and the street, and ended up in that place, and he was not yet thirty.

They came and went, and I was not quite twelve. Whatever passed away would reappear, and what was to come would not be unknown.

Screams, voices, buses rushing by, taxis stopping and speeding off, small trees thrown down in the middle of the bridge, buildings, structures with dirty windows. My mother hated smudged glass. Faces, statures, clothes, trousers, cloaks and red fezzes, black tarbooshes, headcloths that protected everyone from the lethal heat of Baghdad.

The smell of sweat, of rank armpits. The sounds of coughing and blowing snot, of belching and spitting.

A man was pissing against a wall. I surprised them and turned; I stopped and looked.

Your glances were not vulgar. It allowed you to expand the imagination as you accepted the rest: a man pissing, standing still, the wall before him, and all humanity behind him. His legs were apart. His trousers were old, and from between his parted thighs his urine spouted out onto the ground. The stream splashed on the asphalt, yellow, with a sound like radio static. You passed the human urine, you passed the parted thighs, and before he closed the fly of his trousers, he turned to you; he smiled and shook his head.

When your father came out of the bathroom, he drew a large towel around his middle. It reached from his waist to just below his knees. My grandmother sat me down before her and began her rich storytelling. She talked but I paid no attention to her. I was restless and wanted to stand in front of the window. He might pass or stroll slowly by; the towel might fall off, and you would see what my father always kept covered. But Adil was holding my hand, and I did not succeed in seeing anything. He had alertly and securely fastened his trousers and pulled them up higher than usual when we went out. When he slept, he was draped in sheets. When he got up, he stayed with my grandmother and mother; they were the ones who washed him; it was they who made the first inspection.

There was long Rashid Street, broader and cleaner than our street. Its grey concrete lampposts were blotched with dirt, the glass lamps were filthy, the light pallid. This was the Rasafa side; between Rasafa and Karkh, Harun al-Rashid used to listen to riddles and puzzles.

So this was Baghdad, the city of cities. I raised my arm and waved briefly to my mother. She never went to the market. They bound her to the al-A'dhamiyya district. She stumbled there. She came from Aleppo, married in Karbala, got pregnant on a cold iron bed, coughed in the ancient bathroom, and gave birth to us on the floor. Grandmother insisted she went out, always repeating, "Take the children and get away from me. I want to be alone."

"Mama, go with us, have a look at the market, get some fresh air! Aren't you tired of being in the house?"

My mother did not reply or offer any resistance; she went into the kitchen. There she furnished all Baghdad on her table and cooked it at her leisure. She dreamed of it, kissed it, and presented it to herself, made us biscuits, sprinkled them with sugar, almonds, and raisins, and went into a fit of weeping. She alone cried when we left. She had Baghdad bathe with her, and spoke to it in the only room that she knew. The house she tidied, that kitchen whose doors she opened up before her. She washed the dishes and got them as shiny as her eye, leaving her smell in the spoons. She called out, she aged, and when we came back, we saw Baghdad in her eyelids. She surrounded us with her wrist and forearm. We fled from her and she was silent. Baghdad, my mother is the most beautiful thing you have.

The sounds of hammers in the coppersmiths' market, the melodies, the blows against pounded red metal. They hammered melodiously, smoothed out, and balanced. The large, flat sheets of metal folded and curved. Every movement of these strong hands produced something, made an object: a large platter, a basin, an old-style coffee pot, a pan. Their hands lifted up the sheets, the big metal-cutting scissors, the high, narrow anvil which the coppersmith put in front of him; he begins to hammer. The wide wooden anvil, narrow in the middle, was for shaping the metal and adding the engraving and ornamentation, pictures, inscriptions and Qu'ranic verses.

The fire softened the metals and burned our hearts and the muscles of the men competing with the muscular metals.

The young workers and the old men wore unbleached cotton clothes. Their feet were bare. The shops were small. This market was roofed with thin metal sheets. The sound of the hammers grew louder, and the red and yellow shapes changed and evolved. I did not hear the voice of either aunt. Here I can shout as I please; I can sing, joke, stand or walk, or lag behind.

The languages intermingled; everything assailed anything. The cries of the pedlars flowed over me. The ground was furrowed and muddy, not paved. Every moment brought fountains of flame from the openings of the shops, a stifling flame that spread, along with its blaze, black clouds and a penetrating smell. I did not know what it reminded me of.

The paths opened up before me. To the right was a short, twisting, dead-end lane that reminded me of the lane in our neighbourhood. To the left was an open ditch full of leaden-coloured water. The ladies' cloaks were before me, the sighs of admiration behind me. The men smacked their lips at these expanses of veiled women and dreamt of the concealed. I never tired of this sight.

We stood and walked on. Aunt Naima bargained, leaving my father's sister no opportunity to speak. She let her only open her bag and pay out the money.

In this new wedding basin Farida would wash away her first blood, and tall white candles would be set in the middle of the new copper Prophet Zacharias dishes, inlaid in the middle and engraved along the edges. We would put sweets there too, under which we would set out green leaves for the prophet's blessing. Utensils and pitchers, plates and coffee cups, rugs and carpets, silver necklaces and rusty rings. My father's sister poured out her new scent. She did not smell; she had not seen all this before. She took out the money and counted it. This was Grandmother's money – she had prayed over it before going out, to give it a blessing. It was buried in the old bag, intended for today.

We entered the cloth market. Rolls of material, colours, golden thread, silver wires, silken ribbons, and black woollen shawls. The smell of the cloth made me dizzy. Open rose blossoms, the closed circles, the shops turned inside out . . . the silk ornamenting the bride and the lace enveloping her.

Farida's voice emerged from all this commotion, rising and falling as she waved the old jacket around, trying it on me: "Stand up straight! God help your poor teachers at school!" They collected clothes for me from family and relations far and near. They washed and dyed them, tapering the ones that were too short and shortening the ones that were too long. They stitched the back and shoulders to make them fit my narrow frame. They changed the leather on the old shoes and dyed them, and pounded flat the nails that pricked me. The day winter arrived, they took down the velvet curtains and Umm Suturi took them to make them into a coat you would wear for two years, three, it was the colour of the sky when it rains, neither grey nor blue. You wore it in the dirt and mud. You went to Hubi, where it got stained with grease and blood, and to the baker's, where it got sullied with flour and bread dough.

Baghdad's cold paralysed the bones; Mahmoud's warmth spread as he stood before me in his old jacket.

He brushed the dust off my coat and blew it into the public street, and in front of everybody he continued to watch the dust fly into the sky: "When I grow up, I'm going to put these clothes in the museum so people can see our clothes. Huda, some day we'll wear new clothes and read pretty books. My father says things will change, but I don't know how. Study hard so that you'll be a doctor or engineer, but I'm very afraid of blood. You remember when we spat in the street to see the blood? That time I was frightened, as if I'd seen all my blood run out in front of me, as though the street was all blood. I was afraid and ran to the river bank and went swimming, and dunked my head under the water. I swam until I was tired. The sun was strong, and my eyes could hardly see. The other children went away and left me all alone. They said, 'Mahmoud has gone mad,' but no one knew what had happened to me. Whenever I lifted my head out of the river I saw blood, so I went under again until it went away. I don't know what happened then, but I was in the house, shouting and crying, Huda. Lots of blood is scary."

8

This was the day my aunt's blood would be spilled.

Farida was the first fortress of this house. My grandmother and her masters were tested by her: Jamil, Munir, Adil, and I. A battalion that emitted a secret life. She had no double wings, but she did have a skull like my father's pistol, a body as strong as all the men in our neighbourhood, and a voice I heard at night that could scare away the angels. This was her imperial, sublime day. Her secrets would be pierced; everything was upside down. She was stretched out on the carpet in our room, nearly naked, her legs open. Umm Suturi lifted her right leg, and Aunt Naima took her left leg. They drew thick lines along each thigh, clipping and up-rooting the hair of her thighs and legs. She was like steel, turning on to her stomach, her hair hanging between her eyes. The white cube of chalk moved along her brown flesh, swelling, red, rushing into her cells, seeping into her blood.

Farida's voice created a new layer: "Ouch! Let me rest a little while. I'm dying!" Her bursts of trilling were like a declaration of war. Umm

Suturi's voice sounded: "God's blessings upon Muhammad, flesh, fat, and beauty. God help you, sweet Farida. We're almost done."

Aunt Naima: "Listen, Farida, men don't like hairy women. You have very little hair. Just be calm until we finish. For two months you'll see the water running over your body like silk."

She took her arm and got as far as her armpit. The bride wept and fell back.

"You'll be numb any minute now."

The bride kicked among their arms, bracing herself against the wall. Umm Suturi took her head and held it steady in her hands, and presented her face to Aunt Naima. The slender thread moved along her cheek, her chin, and her face.

"Leave my eyebrows!" she cried.

"Today you just be quiet."

Farida looked like a ripe fig. Her cheeks were spotted, her forehead was a blazing red, and her eyebrows were reduced by half. Her beauty was turned upside down. They put her hair up with a narrow ribbon, and her neck turned and moved. My two aunts went down to her chest and stopped at her nipples and the fiery divide. With the tweezers they plucked out her downy blonde fuzz. They palpated her and watched her. They drew flared rays from her belly, which folded over a little, running downward to the last dream, and up to her nipples.

"Look at Huda here with us. Go away. Go outside."

Aunt La'iqa's voice: "When you grow up we'll marry you off ourselves, and rejoice with you. Now go play with the children."

The children, the little ones, and the house was inundated with women. I had not seen them before. I knew Rasmiya, Umm Suturi, Umm Mahmoud, Umm Hashim, Umm Ghanim, and the wives of our neighbourhood's shopkeepers. They all flocked together to the bath in our house, washing and drying, changing clothes, combing their hair, exhaling pleasant odours and cheap perfume, sudden voices, murmuring among the aromatics, slanders coming from their gums: "Yes, this Munir is her cousin. His looks! He frightens the children. But he's terribly rich, and all the luck in the world. Everyone has tried asking for Farida's hand, but life's like that."

"Well, that's her luck. Yes, she's pretty, but she's vain and sour, and her tongue cuts like a saw. When she opens her mouth about anyone, God help them!"

The flood moved into the kitchen as well. Aunt Bahija stood in front of the pots and rolled up her sleeves. She skimmed the grease and sprinkled cumin on the meat and rice. She fried the meatballs. Her pelvis shook as she kneaded the dough for the stuffed pastry. She stirred the pot of milk with her long wooden spoon, blew on the froth, and sprinkled saffron and aromatic herbs on top of the custard.

Aunt Najiya beside her brewed the tea, cleaned the tea glasses, spoons, saucers, and knives. Her voice dripped with greed and jealousy: "Who invited Naima? It must have been you. Are you trying to kill me? Don't you have enough women? Oh, if I could have, I'd have killed you and been saved from you."

The grease, the smells of the kitchen, the distinct smell of Aunt Bahija's repressed laughter. The smell of fingernails painted with grease. The long nightgowns did not emerge from their steam without this panting. The fragrant steam of delicious food warmed the kitchen. The legs of the wedding lamb had been cut, its intestines removed, and the carcass skinned. Hubi had sized it up and taken his share – the head – in advance. My grandmother's voice: "Leave a little meat for us to distribute to the poor at the mosque. This day is a favourite of God's."

It all collided and mingled together, the flesh of the lambs and the flesh of all these women, my mother's sisters Nahida and Zubayda.

The young boys and the girls watched, moving from room to room, not knowing what to do. Grandmother Wafiqa came and went amidst them all. She had bathed before them and still had not changed her clothes. She brought the new silk robe, embroidered with violet thread. She adored this colour and always pronounced it incorrectly. She smacked us when we laughed at her inability to say *banafsaji*, saying instead *banoonsaji*, meaning my teeth have gone and my tongue is crooked through grief.

On this day they called her "the mother of the groom" and Umm Jamil. I pulled her by her dress: "I want my mother. When will she come? Today is my aunt's wedding." She did not reply. Adil was dressed in his old holy day clothes: long trousers the colour of dirty sugar, a blue shirt, and new shoes. He combed his hair and stood bewildered at the gate to the house. He spoke to no one and made no jokes. When he grew tired, he sat on the bench, which the women had pushed out of the way so they might pass without bumping into it.

Firdous was with me as I alternated between the kitchen and the trays of doughnuts and baklava.

The house was upside down. We brought bamboo chairs from our grandfather's big house and the neighbours' houses. These were lined up along the left, and a small table was set before each chair. New mats were laid on the floor, to the right, and freshly fluffed mattresses made up with clean sheets. Big cushions were propped up against the walls. The glasses gleamed, and the dishes and spoons were taken out of the boxes on the top shelf, and were all set out on a rectangular table at the opening of the hall.

This was the wedding of the first girl. My grandmother had lost three girls and a boy; there was only Jamil and Farida, with fifteen years between them.

I went up to the roof and murmured with my father's first voice. Today neither Mr Jamil attended, nor my aunts Widad and Inaam.

I plunged alone into those walls that had been demolished. The storm had harvested all that the other rooms had abandoned. The roof, which had been a bridge leading me to the sky, was now crowded with new fears. Uncle Munir had levelled the surface, paved its rutted mud with coloured tiles, and coloured its walls a bright shade of apricot. The old things were gone; Munir Effendi had taken them out of their boxes and the secret rooms and built the wedding chamber.

He had bought a great wide bed. Its frame was gleaming brass, with scalloped designs at the top and middle. New cotton was brought and the old was fluffed up. A ribbed, bright yellow eiderdown was made to cover it. He bought a wardrobe with four oaken doors, a chest of drawers with a round mirror mounted on top, and a small, low chair upholstered in embroidered black velvet. He replaced the glass in the ceiling with new, coloured glass decorated with a branch pattern.

On the bed they spread out a pink, low-necked nightdress and a robe whose bosom and sleeves were worked in lace and silk thread. The bride's high-heeled shoes were set out on the floor, high and shining with broad feathers. I opened the wardrobe and looked at the bride's clothes. The bundles were lined up: these were her clothes for the first seven days, perfumes, rosewater, sprigs of lavender, dried flowers placed among the bundles, and everything bright, shining, and orderly: blue, red, pink, yellow. I opened them and touched, sighed, longed. This was the corner for Rachel's clothes, this was Aunt Naima's pile. Between them hung Uncle Munir's suits, upright, elegant, new, ironed like a border guard's. Iqbal's clothes had been stored since her trip in a small suitcase tossed on top of the wardrobe. Her first wedding dress, her only silk handkerchief, her hairband, and a picture of her with my father on her wedding day, which she had placed before me on the shelf near the Qu'ran.

This was the new bath on the roof. The silvery pitcher Aunt Naima had bought, the first washbasin, shining taps and door-knobs. A set of new sofas was set up in the other room, all with broad wooden armrests. The women came up here, hung the velvet curtains, set out sticks of incense, and lit them at night. The odour reminded me of the Abu Hanifa Mosque on the nights of the great holy days.

My mother's coughing was absent on this day. The roof swayed once, but no one came up from downstairs and no one asked about the absent woman. From the glass of the roof, I looked down at everyone. Now the roof belonged to my aunt and Mr Munir, and I no longer owned an inch of this sky. And the ground – it was just another form of fever and coughing.

Firdous's voice behind me: "What are you doing here alone?"

She was staring. "God, Aunt Farida's trousseau is beautiful!"

She opened the wardrobe, looked, and sighed. "Quiet and sad. I know you're remembering your mother. Come, let's go down. Pretty soon your aunt will coming out to the house. Come see her in her wedding dress – she doesn't look like the same aunt! Please, Huda. God is good. Your mother will come back safely."

Going down the stairs, she whispered in my ear: "Mahmoud is standing by the gate of the house. He wants to see you." He, too, was wearing his school trousers and holy day jacket, and leather shoes with his clean toes sticking out. His uncombed hair was the colour of cooked turmeric. He had left his bed and the fever. I stood before him on the stone steps. Mahmoud had grown. Now you only looked at one another. The promises of the first laugh were gone. The school's bench was peeling and the flowers in the garden where you both got lost were crushed. Offer him baklava; let him taste Farida's wedding. Share with him the sugar that's dipped in dreams. Put Jamil on the ground and spatter him with the blood of your absent mother.

I had not hated my father enough until now. He did not come. He grew angry and quarrelsome, argued and threatened; he would not consent to this marriage. He stayed far away, but sent a message: "In a few months Farida will be divorced and sitting in front of you again! Munir's real home is in bars and with whores!"

My grandmother did not listen, and did not stop my aunt. My father's voice rang from room to room and in my eardrums. No one heard it but I. I gathered it in my head and approached it. The first time that voice blazed I followed it.

But still the house banished his voice. The nights approached. Baghdad rolled up its sleeves and boiled, and Mahmoud was still boiling before me. Adil's voice: "Dear Huda, I want some sweets." I have a plate of sweets in my hand. We are standing, Mahmoud is ill, good-looking, clean. "Why did you get out of bed?"

He did not reply. My hand touched his fingers. Silence and sugar syrup and the stupidity of people passing by.

The women came in, and among the silence the stickiness ran down.

"Congratulations, Huda, God bless your Aunt Farida."

"Mahmoud – you! How are you now?"

"A little better."

"But your hand is hot and your face is red."

Adil turned. He loved Mahmoud and this sister of his.

"You're as pretty as a rose today."

I would not bow my head. I will always remember this date.

"Your clothes are beautiful, and your hair – when are you going to let your braids down?"

Speak Mahmoud, write, announce, rejoice and don't hold back. I heard Mahmoud's mother's voice behind me:

"Who got you out of bed? Do you want to die and kill me too?"

I left, the baklava in my hands and mouth. Adil grasped Mahmoud's hand: "Go, I'll go with you, I don't like being here alone."

The two shadows moved away. The call to evening prayers could be heard from the Abu Hanifa Mosque. The incense rose from the recesses of the house, and rosewater was sprinkled on everyone's faces. Prayers were read. The bride set out. Two highbacked chairs were put in the house. Palm branches were stuffed into big tubs and arranged amid the tall white candles, whose wicks, once lit, projected high flames. The little candle holders were on the steps, in the hallways, and the entrances to the rooms. Mint leaves, cardamom seeds, sugar ground with nuts were placed in small gilded glasses. Sweets were thrown at everyone. Umm Suturi's voice trilled, and I watched the movement of her tongue and my aunt's head in its first practice: she was the bride; she was Farida. My grandmother called her in a loud voice which everyone present heard:

"Lord, keep her feet on the path to goodness and comfort. Lord, make her happy and spare me long enough to see her children. Almighty God, blessed be the name of Your messenger Muhammad. Darling, trill with joy; all my dears, where are your voices?

"Umm Suturi, Bahija, Naima, darling Umm Mahmoud, God willing when Iqbal comes back from her trip we shall have a proper wedding."

The trilling and voices, the commotion, the prayers sprinkled over them all. The shut-in women of the neighbourhood stood and watched from the corners, looking on and resting for a moment. They were unveiled and coloured; their faces painted with red and blue and adorned with gold taken from its boxes. They wore robes and gorgeous Iraqi over-dresses, worked in gold and small jewels.

Najiya coughed and trilled. Bahija shook her middle and, taking my grandmother's hand, danced amidst the din of the women and children. My aunt was ready to collapse. There were a few metres between her room and the courtyard, and she crossed it in minutes. Everyone was a watcher, wanting to see this ambiguity in the face of brides.

Farida was tinted: her face gleamed, and her eyebrows were more fully arched. But she looked ugly! Her teeth shone white, her fine skin was radiant, her lips were the colour of a new beet. Her finger bore the gold wedding ring, and another finger an emerald-coloured one. On her chest lay a pearl necklace, a gift from my grandmother; the pearl earrings were a gift from Aunt Bahija. Her rings and bracelets, and necklaces were from her sisters and sisters-in-law, Naima, Zubayda,

Nahida; the buttons on her white dress twinkled, and around her waist was wrapped a wide belt that hung as low as her haunches.

The wedding tiara on her head was entwined with small artificial roses. The veil flowed over her neck, shoulders, and arms, and down her back. She did not know what to do with her hands. Now she raised them; now she laid them down. On her right was Aunt Najiya, to her left Aunt Bahija, and before them Naima kept the path clear and quietly watched the two aunts. The rest of the women walked behind them, stopping or slowing down, reciting Qu'ranic verses and trilling, shooing the children out of the way, bearing cushions and trays of sweets.

So this was the bride. She looked nothing like the original Farida. I did not like her this way.

I stood near a tray of candles. My aunts and grandmother called out to me: "Come take the tiara from your aunt. Look – even on a wedding day she's stubborn."

I did not move. I looked down at everyone from the top of the stairs. Beside me, Firdous took a step. I took morsels of the dripping wax and made little balls with faces. I looked up. My mother was coming, wearing a long white silk dress. Her bosom was round and prominent, and her height was exaggerated; she was like a goddess fleeing the earth. She did not turn or speak. She looked only at me as if intoxicated, walking and acting like a woman who knew her private fortune. She walked toward me and touched my face, held my hair in her palms, turned with me, held my hands, pressed them and lifted them up, kissed them and smelled them, and held them to her cheeks, which had grown more plump and healthier, more pink and glowing. Her hair was longer than mine, hanging loose, combed and shiny, clean, and parted in the middle. She wore glittering diamond earrings, which moved whenever she moved. I turned with her; I danced, and we danced. We opened the rooms in the house one by one. We opened the closets, the drawers, the suitcases, the bundles, the boxes, and pulled out all the contents. We opened the windows, walked on tiptoe and sang as if meeting each other for the first time. Our voices rang out and we paid attention and called out to no one else. We did not even recognize the people we knew. She held me by the waist and I embraced her arms; I grew tall, we stretched and grew bigger. Now we had wings, and the great and small houses opened up to us. We did not repeat the same songs or remember what we said. Everything came out of us spontaneously, as if the words knew their cue. She carried me off like a fabulous roc bird, from the house, the lanes, and our street. She ascended and I ascended. I flew and pressed my face against hers, and her eyes saw me as if for the first time. My mother's eyes had grown as large as the ceiling as she slipped out of my arm. I did not see her or pursue her.

Suddenly silence fell. Munir cleared his throat. The trilling grew louder, but with one movement of his hand he silenced them all. He called for my grandmother, and they both entered the room. Only I could hear her cry out: "Oh, God, she's gone, alas, poor Iqbal! Almighty God, I will not resist your wisdom, most merciful of all the merciful."

"Tell me, Grandmother, is it true my mother is gone?"

Look and let her go. Slam the doors behind you and open the windows of all the houses before you. You piss on the cushions and the gold, on Munir's baldness and Aunt Farida's rear end. You trample the mats and the carpets. If only I could have screamed. It was the twenty-seventh day of the month of Ramadan. "A wedding on this day is a blessing," my grandmother said.

And a death on this day?

Today the new visitor, Iqbal, is "gone", today the holy little hairs emerge from her precious flask. God has abandoned me.

Only the prophet is left. Cry out all the curses you have learned by heart in the burial shrine of Naaman Ibn Thabit. Release your amplified scream everywhere, and lift your mother's coffin, containing the future days of both of you.

9

To the Abu Hanifa Mosque. I ran, fell down, picked myself up, cried, and struck my face with my hand. I was thinking of no one.

I stood with the crowd, the drowsy people of our neighbourhood. The long rows pushed and shoved, people bending, their arms and legs. Their voices mingled together in prayer and supplication: "God, have mercy," they cried as they revolved around the tomb with its silver dome and scalloped columns. They wandered and trembled. There were children and elderly people, tying green and white scraps of paper to the window of the tomb. Their hands were like flowers scattered by a storm. Mothers and grandmothers lifted up their little boys and girls, and perched them upon their shoulders. They kissed the columns as if they were suckling breasts; their whole bodies craved the blessing. Their voices were hoarse, heavy, delicate and helpless. Their black cloaks undulated over their statures, rose and fell, and returned to their passionate heads. Handkerchiefs, covers, and towels were drawn over their sweaty necks. There was incense I had never smelled before that reminded me of penitence. I did not know where it had been placed, but it stole through the confines of the space and drifted by our noses like the drowsiness of dawn.

Their faces glistened with sweat, fatigue, and prayer. Their backs bent over, then straightened up. Their steps were uncertain and shaky, and their knees knocked together. The worn-out mats and tattered carpets got shoved aside, revealing stained but cool and beautiful tiles.

Their feet were bare, their toes slender, their eyes swollen. Their nails were long and dirty. The space was wide and vast, dozens of times bigger than our house, and in each of the four corners gilded yellow candles burned. They were tall and thick, like the palm trees we played around on holidays, and their high flames rippled every few moments, whenever the ranks of people passed before them. They flared up and the melted wax ran down through the cracks.

The women thronged together, waiting for Mr Aziz who served as the Abu Hanifa Mosque's administrator: the *mitwalli*. The doors to this place were crammed with mothers, widows, and sisters. I cried amongst them: "Mama, Mama," among the loud prayers to clear the way for the *mitwalli*.

The night of the twenty-seventh of Ramadan, our whole neighbourhood took refuge here. They knelt before the dome and dreamed, stayed awake, listened to the prayers, exulting with them and waiting for the Prophet's hair, hidden all year long in the small room, in the golden casket, in the clean and pure place, by the burial plot behind the mosque. Today it came out of its hiding place, wrapped in a length of thick, perfumed green woollen cloth. The hairs floated in an elongated bottle, whose glass was flattened and slightly bumpy, swimming in rosewater.

Hundreds of hands wanted to kiss and touch it. All these heads glowed and made noise. They bit their lips with joy: "God bless the Prophet Muhammad."

Their bodies twinkled, and the whole multi-layered throng uttered the Prophet's name in loud reverberation, secretive and pure, that rose from their hearts and permeated the entire district of al-A'dhamiyya.

Mothers carried their children into the odours, reaching Mr Aziz's palm.

I shoved and pushed, slid through and jumped between the rows of people: "Hajj Aziz, I am coming through! I want to kiss the hair!"

I was carried by the crowd and pushed far back. I was suffocating, striking out at everything about me and before me. I turned about and cried out, "God keep you, hajji." I clamped down on my braids with my front teeth, and bit down hard on them. I pulled and pushed, came and went, like a maddened wave. My clothes rode up and descended again. I was thirsty. I cried out, "Hajji, my mother is gone!"

Above me, the voices intermingled and flew about. "Your intercession, Prophet of God!" Hands and bodies pushed me, and I crumpled to the ground before the Hajj Aziz. I pulled at the *mitwalli*'s new belted robe and brown cloak with my hands. I turned my face up to him. His face was middle-aged and sad; his beard was uncut, and his eyes as white as my grandmother's skin. His voice was moist: "Go – kiss and pray, because today all prayers are answered. God protect and guide you on the right path."

I was hypnotized. The smells and the moans, the prayers and supplications, and the unheard weeping. The scrap of green cloth had a strange smell. I clutched it to my face, touched the bottle, and he laid it on my head and ran it along my hair. I rubbed it and kissed it. The man's hand was strong, his palm was wide, and his fingers were creased. I turned and shouted, and the voices about me pushed me: "God bless the Prophet Muhammad." Weeping draws me into their protection: "Mama, Mama, where have you gone?"

The voices: "That's enough crying, we want to kiss and smell it, too. Good God, even in this holy place, the greed you see!"

The prayers pushed me away and flung me back. I had asked nothing on this day, I had trusted no one, I had not enshrouded my mother. Her corpse was stretched out before me without limbs or feet. It was just a featureless head.

I looked at it and moved forward on the mosque's tiles. Her face glowed without a shroud, perfumed and cryptic. It collided with no one and contrived for itself only this integrity.

I wiped my nose on the hem of my dress, wiped my tears, sat down and rested my back against the wall. The mats were dry and harsh. The voices of the rows of people faded into the distance. I saw the remains, creased paper, ribbons fallen from my braids, my rusty hairpins, buttons, green leaves, wilted roses, and small-denomination coins.

The old women sprawled out on the floor near me, wiping their eyes and cheeks, and readjusting their head coverings. They sighed. I turned my face, my head, my body to the wall. I stretched out and pulled my hair over my eyes. I had no time to close my eyes. Think of what Mahmoud said; prepare for calamity, and don't return to the house. Sit here; escape from there; sleep here. Kick your house and the people in it. Let your family follow your tracks; let them go out, alone and in groups, carrying torches or sticks. If you saw any one of them now you would shed their blood.

The voice behind me: "Look – isn't that Huda, Jamil's daughter, if I'm not mistaken?" The other woman replied, "Dear, you are Huda, aren't you? Why are you sleeping all alone here? Today is your aunt's wedding. Hmm? Where is your family? Strange, leaving a little girl in the mosque by herself!" If the wedding were ruined, or the bride were dead; had the procession gone off looking for you, or had all the sewers and cesspools opened up; had my grandmother gone to pray, or Adil fallen into a daze, or Mahmoud returned to his fever, or your father passed to madness; if the Tigris were to flood or the king to die – if, if – you would still never return.

Among the voices is Rasmiya's. If she knew, if she saw, if she came, there would be no way out of it: "You are here, Huda? The whole house is upside down over you! Oh, God help you, you're practically an orphan. Wait for me – I want to make my visit here, and we'll go back home."

I turned my head to her, as she disappeared. Her whole form was covered in her black cloak. Her voice sounded portentous in my ears; I could hear the slaps and lashes whistling through the air. I could hear the husband's whistle.

"What are you doing with the needle money? Wakefulness and spirit and cotton; you deliver and abort babies and every time I open the bag it's empty. Hmm? Say something. Where are you taking the money? To your family, your brother the pimp? Hah?"

Her voice rose in his face: "Listen, I've been patient with you. You're the one who steals our money and loses it gambling. Your salary's gone the first of the month. I pay all the debts. The whole neighbourhood knows you're a bully and a sinner."

Rasmiya did not cry. No one saw her tears, but everyone saw her bruised face, her swollen nose, and her trembling hands.

Whenever I ran away from school I stood at the gate of her house, which was always open. I sat on the stone steps and watched the people passing. The wind fluttered the curtain in the doorway in front of my nose, and I smelled the odour of cooked fat and saw the big holes in it. Her house was near ours. When Abu Ayman came out he saw me in front of him: "Hah, Huda, aren't you going to school today?"

"We were let out today. Our teacher is ill."

When he had gone away, she turned on the radio and began cleaning up inside. Rasmiya put up with the beating and sang. Her voice reached me, and I sang along softly. She put rings with big and showy jewels on her short fingers. Thick gold bangles glittered on her fat, hairy forearm, dense, brilliant, and crowded together. When she went up to wash the outdoor steps, she stood before me, a bucket of water in her hand, murmuring slowly as she stood: "Do you run away from school every day?"

Her nightgown exhaled the odours of sweat, cooked rice, and her stifled tears. Her breasts were on the verge of sticking into the whole street, and her black hair was pulled up into a cheap red handkerchief. Her neck was short, her arms bare, and her shrill laughter revealed a gold tooth in her upper jaw. When I looked up into her face she did not look at me; she continued singing, her face sallow, a light blue tattoo on her jaw. Her eyes were narrow and black, her eyelashes thick, and her body tense. She still held the bucket. I stood up.

"Good morning, Aunt Rasmiya."

"Shame on the books and notebooks. You say you left before the lesson, hmm?"

I did not reply. I heard her harsh breathing as she saw Abu Mahmoud, the cheese seller, in front of her. His shop faced her house. He wiped his face, straightened his cap, and hitched up his old striped wide-belted trousers, shooed the flies away from the cheese, and lifted a morsel of white cheese to his mouth. They looked at one another; the crevice of her breasts was in front of him. He spoke: "Arab cheese, Kurdish cheese – if you don't buy, just taste."

She laughed sweetly and easily, and poured the water out, drop by drop. She bent over, the bucket in her hands. Half her height was before me, and his whole face was before her. He smiled and nibbled the cheese, lifted the palm frond and arranged the chunks of cheese, now far, now near. The water followed me, its spatter and its sound. Passers-by stopped at his shop to buy. He sold cheese, cut, weighed, and wrapped it. He wiped his mouth with his hand. He went over to her; she stood before him. The water was gone. The steps were clean, and the radio was still on. And her chest . . . my schoolbooks fell to the ground.

Every day I ran along the high wall of the roof and ended up in the bath. My aunt washed my face, combed my hair, and tied the ribbons on my braids. She dressed me in the dark blue school smock, rolled me up before her and held me by the arm. She ran after me and I dodged her, hiding from her voice and her shouts. She always found me and led me by the hand to the entrance of the school. She moved away only after I was out of her sight. Before leaving the house, I stole five fils from my grandmother and pressed them into the hand of the school janitor, Abu Muhammad. He opened the gate for me, and I passed into the street after the third lesson of the day.

My aunt's voice split the air as I wound my way through the street: "I've whipped you, but even beatings don't do you any good. Do you want to leave school to drive me completely mad?"

I walked around the Naaman Park, gazed at the earth dam, the Tigris, and the fishermen. I went down to the al-A'dhamiyya Park and waited for Firdous and Adil there. I read the advertisements for Arabic films: the face of the Egyptian actress Fatin Hamama covering the whole wall. It was a sad face, as ill as my mother's when she'd had no injection and taken no medicine. And Farid al-Atrash standing behind her, as fractured as the ice we put in the glasses of homemade fruit punch.

"He looks like he's going to puke," said Firdous. "I don't know how the girls in school can love him."

Adil smiled and I cackled. Mahmoud emerged from the short trees in front of us:

"You skipped school today too?"

I did not reply. We walked along together. Mahmoud stared into the distance, not blinking or flinching. His school was far away, his trousers were long, and his thighs were big. The school always surrounded you. Its walls were high and its classrooms dim and small, their dark paint peeling. The benches were narrow, the blackboards smudged. The boys and girls were crammed together in them, four by four. Miss Karima wrote the dates of wars as forcefully as the way my father hit me. She spat on the floor and sat facing us, her legs apart. Her belly was distended; she was expecting her first child. When it moved, she yelped and put her hand on her belly: "I want a boy – otherwise I'm going home to my family." She rose and moved slowly and almost fell. We grasped her by the arm, but before she got to the door she began to vomit. She left and did not come back until after the birth of her son.

Miss Bahira shouted at us about subjects and predicates, verbs and subjects of verbal clauses. I always laughed in her class, repeating her sentences after her. She grew angry and her colour changed. She snatched her ruler and smacked my fingers as I counted the minutes and hours in order to put Arabic grammar in the frying pans, and let them burn on the fire.

I did not stop until getting my bitter punishment: fifty lines of grammatical sentences. When Miss Qadriya came into the classroom she immediately reminded me of my mother. She was as slender as she was, with unforgettably large white, symmetrical teeth which were always clean. It was as if she had never used them for eating. Her eyes were big and bulged, and her hair was long and the colour of molasses, though she always wore a white kerchief over her head. They called her Haja Qadriya. She listed diseases, discussing prevention and treatment, medicines, vitamins, and mentioned all the diseases, but not pulmonary tuberculosis.

I did mathematics lessons only with Mahmoud. He taught me the rules of division. He understood odd and even numbers, tens and thousands, fractions and decimals. Mahmoud loved numbers. He piloted them, wrestled with them, danced before them, and they bowed to him. When Firdous made a mistake he shouted at her, and when I made a mistake he rounded on me. Patient, stern, he never retreated from a number, and the numbers acknowledged only him. He laughed in front of us, saying:

"These numbers are like people. One time I sat down at night and counted from one to ten. I gave a number to every person I know, and when the numbers ran out, I knew that there are some people who can't be turned into numbers. The number falls in front of them, and they remain decimals or fractions, they don't know this language and don't want to learn. Only I love them, even if they don't understand anything."

Before he finished:

"And me, have you given a number to me?"

He was silent and lifted his head to look at me. His face was flushed. His eyes were the colour of ground coffee, his nose was big and his forehead was high; his cheeks were full and his lips heavy. It was full but melancholy, long, and animated. When he laughed, his front teeth seemed to jut out. "I don't know, little Huda. Read well and you'll pass. Our lessons are silly and the teachers don't teach maths well, but this is all we have."

The day I failed Mahmoud disappeared. His anger did not dissipate for a month: "This year I'll forgive you. Your mother has gone away and your father married, but if you had passed in the middle of all these troubles things would have been very different."

And things were very different. I stood before my aunt, my grandmother, and Adil, and I announced my decision: "I want to sleep and study in my mother's room. I'm grown up now and I want to study by myself. My father no longer comes to see us the way he used to."

"Yes, dear. Anything else?" My aunt's voice virtually sizzled. "Does her Highness have any other orders?"

"I want us to listen to the radio a little. Every day we listen to the Qu'ran and then turn it off. I want to take the radio with me into the room."

My aunt wanted to tear me to pieces.

"Fine. fine. By God, if you were my daughter I'd kill you. I wouldn't let you go to school or see the street again in your life. Oh, God! There is no power or strength save in Almighty God!" I laughed softly. The locked room would be opened, and I would be soaked in my father's sourness and my mother's happiness. I opened the closet and looked, felt, and smelled. I rolled around on the big bed. Sleep here. Sleep backward, look as though you're studying but put thick novels among your books. Roam there alone, perfume yourself with slips of the tongue and sins of the heart. Go from one scene to the next. Catch your breath and open the box of odds and ends. Look at the shelves. All these books will be mine. The spiderwebs are mine, and all the dust of the rebellious nights. Here you may scream at al-Aqqad and Taha Hussein. Turn on Baghdad Radio and listen to the royal anthem. Turn the dial to Voice of the Arabs from Cairo and enter Paradise with the voice of Nasser.

The first night I could not sleep. I was not afraid. No ghosts faltered on their way to me, nor was I afflicted by nostalgia. I was gaining my mother for myself, and gathering the power to challenge my father. I went to Adil and did not escape from my grandmother. My aunt was like the lessons I took at school. When we set off for school Adil was waiting for me, looking sad, just inside the hallway: "I didn't sleep well. Every little while I sat up and saw your empty bed. Huda, what did you do in the room all alone? Were you afraid?"

I ran toward the courtyard of the mosque and sat on the stone wall under the lofty lotus tree. I looked up at the sky and escaped from Rasmiya.

10

⁓◉◉◉⁓

The mosque had no times for women, but the men entered in droves. Grandfathers, uncles, boys, strong men, young and old. Men you knew, and men you did not know. No man lifted your mother's coffin. They yawned at it in Aleppo, coughed, and the night of indifference fell. Iqbal's belongings were divided up amongst everyone.

Mr Jamil did not open the telegram. He set his table and added another glass. He circumnavigated his wife's belly and awaited his third son. Only Munir entered victorious. He headed inside and turned his back on them all.

They came in, men with round bellies and wrinkled paunches, and chests awaiting metres of pardon and health. Their supplications flew like Adil's paper kites in the courtyard of the mosque. They gleamed like crystals. Their sandals were dyed and their shoes shined, their cloaks were ironed and their hair combed, their moustaches perfumed and their head coverings clean. Their wide belts slipped below their bellies; their long robes and long garments were diaphanous; their suits had been pressed in expectation of this day.

By night these men did not resemble the men they were by day. These had come from houses, coffee-houses, and bars. They defused resentment and blame. Everyone waited for everyone. They shook hands, clapped one another on the back, stood up; their creased foreheads met: wide, narrow, haughty, away from the window. Their voices echoed among the silver columns, and they knelt humbly.

You had never before seen a man kneel. Your father never prayed. Adil had practiced a long time but forgotten the first move. Your aunt knelt only to herself. Your mother always prayed alone. You had never seen any of these.

They prostrated themselves, and stood up again. Gestures, prayers, movements. Just as from a distance, you did as the praying women did, up close you prayed as the men before you prayed, their heads and bodies submitting to God and the Prophet and those close to him in the presence of God: "Peace be upon you, and the mercy and blessings of God."

The opening *Sura* of the Qu'ran is on their lips, recited softly in perfect unison. They lift up their palms, wipe their faces, and move, row upon row, waiting for the voice of Mahdi, who knows the Qu'ran by heart, to recite the holy verses. Their black and yellow prayer beads come out and are looped around their hands and fingers.

Some of them leaned against the wall, others on the water tank. Silence fell. They coughed with deliberation, moved their heads, straightened their backs, breathed gently, and listened to the voice of the reader entering through the crevices in the corners, coming through the walls. I recited along with him, walked, and heard him in the wide courtyard of broken flagstones. I stood in the middle. The children and women gathered themselves and their cloaks, peering into the back door of the mosque.

The Abu Hanifa Mosque was the mosque of my first quarrel, and my first temptation to play under the tall lotus tree. One day we sat under it, the boys and girls of the neighbourhood, we laughed, and chased one another. We shook the dry, sour lotus tree and dodged its big, oily fruits. We played hide-and-seek around it, hid our head and sides, and grabbed one another by our shirt-tails, braids, and dishdashas. We encircled the stump with our arms, Mahmoud covering half of it, and I had to make up the other half, so that we were clutching the tree against our chests. We shook it, and our fingertips touched and trembled.

We waited for its short season of fruition. We all lined up underneath it, and Mahmoud climbed up, shook the dense branches, shook them and shook them some more. I looked up and saw him looking down only at me. Firdous shouted, "Mahmoud, be careful!"

She turned to me: "It's all just for your sake, so you can eat some ripe dates!"

I laughed and did not answer her. Everyone gathered the dates from off the ground.

Mahmoud filled his pockets with ripe, juicy dates, so plump that the skin was bursting. He came down, seeing only my arms waiting for him. We held up our skirts for them, counted and ate them, swallowing the pits. We smacked our lips, savouring the taste of dates in our throats and between our teeth, never satisfied.

Here we took the first of our hesitant, unsteady steps, until our feet grew a little heavy. Here we opened our eyelids on the high domes. We ate hot bread and cooked date paste with sesame and nuts. We carried plates of elaborate and expensive wheat-and-meat pudding, pots of stale food and dishes of date molasses and milk.

My grandmother sent me here on holy days. I carried the tray and made the rounds of the old men and women squatting in the corners and on the stone steps of the mosque, who lifted their heads to me: "God bless Umm Jamil and help her prosper. She has never forgotten us."

They took us together. Nahida's girls, Afaf and Ansaf, stood face to face, not moving, not blinking, not extending a hand. They were the intelligent girls, clean and well-mannered. We did not smile or speak, only stared. Suddenly, they hurried off after their mother.

I watched them process around the tomb.

They all came here; my grandfather's big house; the elderly of the neighbourhood; the women from other streets. We went around and around, then stopped before his window, and the voice of my aunt called from behind me: "Pray and ask for what you want. A plea to Abu Hanifa is never ignored." She pushed me in front of her as if I were on my way to school. She surrounded me, and I heard the movement of her cloak and her rapid, heavy breathing. She waited for my tongue to spring into action as she enveloped me under her armpit, the sour smell of the sweat that trickled down her neck and belly:

"Hah – have you gone mute? This is the only place you show any manners. Have you prayed or not yet?"

"Will you pray for me?"

"What's wrong with your tongue? May God cut it out and give us a rest from it! I'm not praying for anyone. Everyone prays for his own soul."

"I don't know how."

"Say, 'God, guide me on the right path, may my father prosper and my mother, brother, and grandmother be blessed with health; and may my aunt marry very soon so that she will be rid of Huda and that damned house.'"

"Why don't you pray for us all?"

"Oh, for God's sake! Even in this sacred place you're still a brat? Where on earth did they get you? Fine, I'll pray by myself. I don't want anyone near me listening to what I'm saying." She pinched me by the forearm and went away from me. I did not know what should be said in front of the tomb. My grandmother sat far off, telling her beads. Her gaze took in none of what was all about her.

She did not pay attention or speak to anyone; she had gone, she was travelling; all that could be seen of her body was the triangle of her face and the shiny glass of her spectacles. My mother always sat beside her; she turned, prayed, and asked for what she wanted. She wept silently and then sat down, waiting for nothing. Adil squatted between both of them, watching everyone, never budging.

I slipped away from my aunt and her voice behind me, and looked out at the courtyard.

These high domes. The main gate of the mosque was to the east, a rectangle of carved yellow bricks, bordered by inscriptions cut into their blue surface. To the north of the door stood the four-sided clock. One clockface counted the hours from sundown, and the three others reckoned the time from noon. They were faced with yellow aluminium. The clock was unique in its form and its beauty. The mosque had two domes and a minaret on the east side, one over the sanctuary of the praying area, and the other over the tomb of Abu Hanifa, may he rest in peace. They began with the renovation of the outer enclosure with yellow brick inlaid with blue tile from Karbala inscribed with some of the ninety-nine beautiful names of God. When I lifted my head up to the blue minaret I heard the sound of the beating of turtledoves' wings and those of grey and white pigeons as they took straws for nestbuilding and flew higher, alighting on the top to sleep there.

Suturi raised birds like these. We used to call him Suturi the bird boy, and stole grain for him and hid stale bread and leftovers for him, and went up to his family's high roof. We cleaned the nests of droppings and threw grain to the birds, soaked the dry bread, and scattered it on the ground.

Adil loved Suturi's birds. He ignored us and went to his house. They stood together, watching the birds flap away:

"Suturi, look at them, look how far away they've flown! I'm afraid they won't come back!"

"No, they know the way back better than you or I do."

"Fine, you know birds better than me. My father won't let me raise birds. He says, 'You have paper kites; they're like birds, fly them and let the birds fly in the sky where they belong.' "

"I don't imprison them. Look – the cage is empty and the birds are in the sky. I just like playing with them. When my father died I began to raise them."

"Yes, I see, but I'm afraid they'll lose their way and not come back, like my mother."

I walked, and mounted the stairs. The air was arid and suffocating. The leaves on the trees were not stirring. The gnats and mosquitoes attacked. They came from the nearby riverbank and the dense trees. They buzzed and droned and bit me, stabbing at my face and arms. I hit out at them and tried to catch some; I pounced at them as they flew into the beam of lights in the corners of the mosque. I scratched my skin, arms, and legs. I ran around the corners, shook down the leaves from the trees, gathered them, crushed them in my hands, and waited for the sap from their dry stems.

No one noticed me. The darkness was spreading over the houses and closed alleys. It was a yellow, plague-stricken Ramadan night. The sound of the big lorries made the street tremble. The holiday sheep were taken off, and the hoarseness of their voices reminded me of my aunt's. They pushed ahead and walked all around one another, waiting to be slaughtered.

My eyelids were festering, my hair was unkempt. Come, Mahmoud, I have let down my braids for you. You want it down loose, and I want it flying, fleeing from the houses, from brothers; come, children of the neighbourhood, I'll follow you from the riverbank to the grave-yard, and from the school to the public parks. Stretch out your palms and wet them with my tears. Light the lamps of the mosque, the houses, coffee-shops, palaces, and the courts.

The men came out, moving, walking, like light, quivering birds. They went to their houses nearby, or dropped by the Naaman Coffee-house, ordering strong tea, listening to the popping of the coals as they burned in the filthy hookahs.

They passed the backgammon board back and forth. The players' shouts grew louder, and smoke drifted from the students smoking and studying at the rear of the coffee-house. The pensioners' coughing mingled with mentions of the names of Nasser and Nuri al-Said, Salih Jabr, the mandate, the beautiful young king and the English, the demonstrations and the leaflets, the government and Voice of the Arabs Radio from Cairo. All these were on the men's lips, softly, in whispers – in fear. Cigarette smoke and coal fumes irritated their eyes. They shouted, murmured, and fell silent as they waited for their plates of skewered kebabs, goat's testicles and livers. They ate and joked, and all left before the call for the last meal before sunrise.

A parasang separated our house from the mosque, with just a few more metres from the coffee-house to the other houses. When the men came out of the coffee-house, the women left their informal parties and faraway alleys. Married women, widows, and young women waiting for a strong back, for a roof that did not leak, and a faithful man. Hand in hand, three, four, they came out and walked down the deserted alleys. They went into the other streets, reaching the dirt dam, and the lights of the lofty houses shone like diamonds. Their cloaks enveloped every inch of their bodies, and veils covered their faces. They did not cough, blow their noses, or sigh.

The game began; it had been agreed upon the previous night. The women I knew brought their offerings from the main street. They watched the needle of the scale and how it moved among the mouths of the men passing before them. The women walked behind them calm to the point of suffocation, waiting for a slip of the tongue from the first man, to give them a way to deliverance and good fortune. A man might say to another, "Today was a good day – I made one sale and got a week's worth of profit. My friend, all is still well with the world."

The other might let him in on his secret: "Oh, God, my wife – she can't get enough gold. Her demands start as soon as I open my eyes in the morning, and we don't get to bed at night until she's got all my money from the day." If the conversation turned to politics, they stuttered a little, and barely audibly: "Yes, my sister's son has leaflets. It's driving her mad."

The women memorized every word, analysed and interpreted, sought out the hidden meanings and insinuations of this unknown creature: man. The whole web moved about him; with obscure codes he was present even in his absence. They followed his star when the sun rose or the crescent moon disappeared. They revealed their inner selves to him and contacted him from within their rooms and cloaks.

That is what my aunts did: Farida, Najiya, Bajija, Rasmiya, Umm Mahmoud, Zubayda, and Umm Suturi. They all imagined a devilish companion giving them a name, a title, and security.

They hid their rusty darts in their hearts as they returned to the humid alley, the stained steps, the cold family, and the constantly nagging children.

Rasmiya's voice rose among the others:

"I saw her sleeping in the mosque, with the women. Where would a girl of her age go on a night like this?" The voices of my beloveds; and they were coming toward me.

They stopped in front of me and I ran away from them. They chased after me. I raced. Hashim grabbed me by my legs instead of my braids. I could not see their faces, and the last pallid rays coming from the shopfronts. Firdous pulled one leg and pushed the other, and suddenly her voice was rough: "Huda, stop a minute. I just want to tell you one thing. Fine, come to our house."

Mahmoud and Nizar tucked me between their arms. Adil was standing, motionless. Everyone stood at my head. We fell to the ground, and I kicked and panted. Adil huddled in my embrace at last, and we shouted with one voice, "Mama! Mama!"

They pulled me away and we dragged our feet. Firdous took me by one arm and Adil seized the other.

Our house was filled with the same people. The chairs had been set out on the roof, and the whole house was carpeted with rugs and thick mats. Not one candle burned. There was no hiding place where we might escape Iqbal's coughing. The dishes had disappeared. The glasses had been put away. The sticks of incense had spread a new odour that reminded me of Suturi's birds.

All your aunts, the women, everyone you knew was in the hallways, in the kitchen. They sat on the stairs. The bride had removed her tiara but kept her wedding dress on. She sat, her legs open, and began to clap. Grandmother leaned her back against the wall, a Qu'ran in her hand. Umm Suturi stood in the middle of the house, in the middle of them all, like a slaughtered bird, her cloak belted around her waist, her head uncovered; her kerchief had fallen off. Her hair was a mingling of dusty red and faded black. The smell of black henna mingled with sweat and animal fat. Her shoulder-length hair was loose and fanned out, and covered her face and eyes. She wailed and lamented, striking her chest.

Her voice shook the house. My aunts stood in the middle, the breasts of their dresses open, their hair loose, their voices splitting the air as they repeated after Umm Suturi. They beat their chests, faces, heads, cheeks and foreheads, with their hands, which rose and fell as one, up and down, as if connected by an invisible thread. The circle grew, and they put Adil and me in the middle. We got down on the floor. They got down in front of us, striking themselves and shouting in our faces. Umm Mahmoud was in front of me. Aunt La'iqa did not weep, but struck her face; her large breasts heaved up and down. Her pelvis ground back and forth in front of me. All the women on their feet twirled around one another, striking their faces in front of one another. They were afraid Aunt Najiya had stopped breathing. She moved her head, but her hand could not reach her open bosom. The women of the neighbourhood were lined up in the hall, crying and covering their heads with their cloaks. Umm Suturi was the toughest of all: she did not cry, but stood and drew us to her voice, which changed and overflowed with sadness. The sobbing intensified as she screamed:

"Don't think I've forgotten you
My notebook's in my heart and I'll write to you
We were forced to part with you."
Their ribs broke, and they all inhaled as one. Everyone repeated, "Alas! Alas!" My grandmother said nothing. She prayed softly, as if standing in the desert seeing my mother before her. She turned the pages calmly. Her face seemed whiter. Unshed tears shone behind her spectacles. Her hand did not tremble. Her chest rose and fell. Her asthma came back whenever there was a wedding or funeral. She coughed and choked. They brought her water and medicine, and she quietened down and her voice sounded as though she were addressing herself: "Lord, You are all-knowing. You are the protector. Oh Lord, take her from my heart as You have taken her from my path. Every day You test me, Lord. Almighty God, I do not oppose Your will. God is great! She is already one week buried. Jamouli knows that but he won't say it. Why? Why? Dear Iqbal, this too is a test. God is constantly testing His servants. Rest there in the gardens of grace. I prayed the Sura of Ya Sin forty times for you, and it will do you some good there in Heaven. It will relieve you of some of the pain of this dirty world. A thousand mercies to your soul!" Whenever she closed the Qu'ran she kissed it, rose slowly, seeing no one, even though their shouts tore the air.

Suddenly she pushed the bodies aside, seized us, pulled us, shoved us before her. No one stood in her way. We stumbled among the cloaks and shoes. She headed upstairs and pushed us out to the roof. She opened the door and stood face to face with the sky. She walked on, with us beside her, holding us tightly. We were trembling all over. She removed her spectacles and placed them on top of the mud wall, then looked at us and pulled us in wordlessly. She took us to her bosom and hugged us, and we buried our faces in her belly, sobbing and hiccupping. Her voice blotted out the whole sky: "We'll cry here – I am with you. Cry for your precious mother. Cry your tears out here. When we go downstairs I don't want to see any tears in your eyes. This is the way the world is. We come and we go. Others come and carry on. No one remains. Even the Prophet, the beloved of God, was taken, to be with Him. Only Almighty God remains."

She took our hands and sat us down before her, crying from the bottom of her heart and lamenting:

"Alas for her who locked the door
And cast away its keys.
Who left; from whom we'll hear no more."

11

ॐ

"Little dove of mine, where have you gone?" You have left.

The sounds of the dove mourning on the domes of the mosques and the distant, lofty treetops where it settled. Coming from far away, distorted, they reached my ears, and I repeated their song after it. We were alone; when we felt lonely we stood on the walls of the roofs, between our house and Mahmoud's. He moved his beak to the rhythm of a nocturnal flute. He murmured when he realized he was still alone. I arched my back, pulled in my stomach, and moved my head and arms, then my fingers, and we shouted aloud together.

When he heard the sound, he began to flutter his short, ruffled wings as they beat against one another and in the sound of flapping grief was gently stirred. It moaned, and I choked back my tears.

My grandmother coughed as she turned on her mattress, then got up and walked to the roof. She looked at our beds and lifted her face to the sky. She murmured, and emitted a hacking, staccato cough. She turned to me but did not see me. After several breaths, she began coughing harder, then tossed away her cigarette and stepped on it.

We came up here, Adil and I, every long, hot summer month. We carried our thick mats up from the little room, aired the sheets and pillows, swept the floor, and sprinkled it with water. The dirt smoothed out, sending up a pale yellow dust that made us sneeze for a long time. We ran around and played, flexed our arms, bent our fingers back. We put out the old chairs and the broken bricks and stood on them to catch a glimpse inside the homes, the roofs of the houses, the dove nests, the bathrooms and their chimneys, and at the colours of the cars and trucks passing by at a distance.

Adil rode the iron beds, jumped up and down on this one and ran circles around another. He looked up at the sky. He jumped in front of me, the pillows in his hand, throwing them at my head one after another and shouting, "Look at the sky! It's the same colour as our grandmother's face. I don't like winter – water leaks through the roof. We can't play in the street because of the mud, and the cold makes Grandmother ill. Look at those birds – when they fly high, they might want us to wait for them. Huda, don't you like birds?"

Grandmother was now standing over me. I looked into her eyes but she could not see me in the dark: I took her hand, and she took mine: "Huda, when did you get up?"

"Whenever you move I hear you. I haven't slept like you."

She sat on the edge of my bed, and laid her hand on my head and face. I kissed her and hugged her hand: "May God give us all patience. My tears are dry – grief will blind me. Today we'll be travelling. In a little while we'll go to the cemetery and then we'll go to Karbala."

"Why Karbala?"

"Today is a holy day. Have you forgotten? We are travelling to see the absent ones – God, I do not oppose your will." She burst into a fit of slow sobs, and then I did as well. She resumed in a grief-choked voice: "Iqbal is gone, and Jamil has forgotten us. He has forgotten his mother, his sister, and his children. God is good." She wiped her tears and went on. "Yes, we will go. I want to weep in the presence of the Lord of Martyrs, Hussein, may God honour his face. We will ask him to soften Jamil's heart and heal him, and I will ask him for patience. If there is enough time, we'll visit Najaf as well." She stood up, repeating, "Lord, strengthen my faith and help me not to complain too much. You are our helper and our protector. Help me to need no one – not

Jamil, not Munir. Lord, may I not seek help from any but You. Take me before my strength and sight are gone." I rose and stood before her; she took me in her arms and stroked my hair.

"Who will go with us?"

"All of us are going."

"Even Uncle Munir?"

"Munir is gone. No one knows where he is. He did not ask, or drop in. And he didn't listen to what people say about his uncle's daughter."

She left me and went away. I saw her like a tidy angel, ill and utterly encircled.

Dawn had begun to pierce the skin of the night sky. The doves fell silent, and the open-eyed crows came. My grandmother always dropped by when we were going to sleep on the roof: "If a crow settles near you head, and caws, it's a bad omen."

Our sky was a dwelling place for all the crows. They nested near our houses, and competed in cawing. They flew so close to my head that I could hear their wings beating. They shrieked and flew higher, and I began to shudder. The crows did not come until Iqbal had gone; only the doves could be heard venturing out of their nests. They began singing to us as soon as we opened our eyes: "Mother, my sweet, my comfort."

"Little dove of mine, where have you gone?"

Farida woke up gloomy, muttering, her voice heavy. She did not say good morning to anyone. I got Adil out of bed and took his hand, and we went downstairs; he was still sleepy. We splashed water on our faces and put on our old clothes. It was still dark. The family left the house. Adil grasped my hand but I slipped away from him. The women of our grandfather's big house were looking out at the edge of the neighbourhood, and the winds from the cloaks newly come from the dyeworks reached our noses. The mothers of Mahmoud, Suturi, Hashim, and Iman were utterly silent. Dawn in Baghdad was like the foam in Umm Suturi's tub: the clouds were enormous, as if wearing grey and black cloaks.

Great Imam Street was quiet. The coffee-houses were locked up, the lanes were secretive, and the rubbish caught my attention. I saw only bags strewn around in front of the steps of the houses and closed shops. Skinny dogs stood bewitched before the steam rising from the bags. Big she-cats fought and licked their wounds, and scattered the whole mess, violated by their claws and spittle, pulling remnants from it into the corners. Grandmother coughed timidly. Her asthma came and went with the rhythm of a pendulum, never varying by an hour.

The dim street-lights made the shadows bigger and longer before us, so that Adil stopped shouting at me as he ran behind me:

"You're quiet like them. Aren't you afraid of the dark? Say something!"

"Keep walking and be quiet."

"Who are we going to see there?"

"No one."

"This is the first time I've gone with you. Is everybody quiet when they go the cemetery? Fine – what if a genie comes at us?"

"We're all with you, don't worry."

"You talk and I'll be quiet."

The walls of the mosque were surrounded by standing figures. Women and men with their arms outstretched. Their sleepy voices rose in prayer. A light, cool breeze brushed my head and rippled the pores of my skin combining with the shaking of Adil's fingers as they gripped my hand:

"Do you remember when you took me to the cemetery? We were young and we played round the tombs. It was hot and sunny, and there were people about, and I laughed to myself and said 'Huda will be afraid and take me home. But, God, we kept on playing. We didn't eat or sleep, we just played. It was fine even if she scared me.' Huda, I'm telling you the truth, don't be cross. That time I wasn't afraid. I wanted to see your fear. Huda, aren't you afraid?"

"You can't stop talking today! Pretty soon the sun will come up and we won't be afraid."

"When either one of us goes quiet, my fear comes back. When people die they go quiet." He trembled more, pulling at my hand. We stopped together and I hugged him, but he did not hug me back. His hand was still gripping mine and his body was shaking, though his tears did not fall.

Men came out of the lanes and intersections, clearing their throats and spitting on the ground as they walked by.

I heard everyone coughing as if they were tuning their vocal cords before entering.

As soon as we set foot in the cemetery we heard verses from the Qu'ran. The reciters sat among the tombs on bamboo chairs. The tombs of people from the other streets brought the reciters. People from the little culs-de-sac recited for themselves, and wept.

The women of our neighbourhood were before me, their eyelids open, gazing at the writings covered with rainwater, soil, and oblivion. Every woman prayed and wailed at the low, compact, covered grave. The young men collected around the graves, looking at their weeping mothers, and wept with them.

The cloaks opened to reveal vast sweaty bodies. Sighs coursed from their chests to their throats. The soil crumbled between their fingers. Persian anthills, large and small, black and red, opened their caves and crept out among our fingers.

The war of lamentation began.

Grandmother's voice gathered strength, bit by bit, as she paced and then sprawled out on the ground. Her tears glistened like stars hoarded in a cave. She prayed, though all we heard were the ends of the words. Then her tears started, quietly and decorously; she wept until the whole grave was covered with tears. She passed through Aleppo and Mecca; she pronounced the Prophet's name as if washing herself with it. She felt her pain light the incense for him, and distributed it equally among all. She did not forget us, who stood round her:

"Pray the *fatiha*. Huda, breathe on your mother's soul. Adil, my boy, don't make any mistakes as you pray; send them to her pure soul. The soul can hear, and feel, and get upset as well. This is where we will all be buried."

In front of the tomb of our grandfather, Ahmad Maarouf, she bowed, murmured, and prayed. No one stood near her in her journey. She looked at the ground as if she wanted to rip it open with her bare hands. We saw the trembling of her fingers, the tapping of her palm and an anonymous lamentation depart from her chest, going with the movement of the waves of the Shatt al-Arab. Their sound seeped into the crevices of the boulders and buoyed the small boats moored to the shore. She passed by the small boats. She was stung as she pursued the large ships. She passed the clumps of palms. She exchanged glances with the upright fronds. There she curled up every day among the soil of the walled gardens, densely packed with trees, hoarse and spent. She cursed calamity and the crows, and waited for the blessing of the water as it spoke to her. She prayed the dawn prayer facing the shore. She picked up some of the salt with her tongue as she trembled in the night. She smoked fifty cigarettes and camped on the riverbank and allowed no one near her. Alone, she opened the line of sand and waited for the man's tie in case he should come up to the roof or a wedding ring should appear on her hand.

Grandfather's boat sank in the Shatt al-Arab, and the Shatt was crowded with the names of the six adventurers. The employees were: the inspector of the administrative district, Mr Ahmad Maarouf, the police delegate, the director of the treasury, the precinct physician, and two guards. Their muscles relaxed in the rise of water, they grew heavy and sank beneath the crabs' legs and the jaws of the electric eels. The bed of water took them, and there they fell asleep forever.

Facing the shore she set up a large tent and said, "Here we will mourn for seven days. Here we will await their bodies." She drew her cloak around her and spread out her carpet on the waves of sand when the tide pulled out. She opened her arms and hugged herself, and went inside when the islands appeared. She walked, wiping her spectacles clean of the sea spray, and went down on her narrow white feet. She had drawn her wedding shift tightly across her middle, and in her hand she held a lantern whose wick flickered whenever the night air stirred. She wailed:

"I have come to you belted with your sash, Abu Jamil! I want to bid you farewell here. You have been gone too long this time. When will you return?"

She waded into the water which radiated from her in successive rings and called: "This is your belt, Abu Jamil. Come and see. It's the first time I've come out to you here. Don't delay, Abu Jamil. Alas, poor Jamil and Farida! Poor Wafiqa, who will come after you!"

She did not grow weary; she did not grow angry. She pursued the water and waited for some sign of the missing man: his striped broadcloth jacket or grey trousers. She waited for her whole life to pass in front of her eyes as she caught sight of a round thing far off, dented, now floating, now sinking, now visible, now not. It was not a person; nor was it any sort of animal. Its colour was somewhere between black and violet. Mysterious, it floated slowly.

She screamed, then fell silent. She paced slowly and moaned. The water flowed over her, and she splashed at it and beat her palms against its surface. Everything round her flapped and fluttered: the birds, the water, the fishermen's boats and their ancient nets. The local people's faces looked on as she stood on the shore. Jamil's voice was nowhere to be heard. Farida stumbled about and cried, wanting her mother. No one interfered with what she was doing; no one was worried about her. They said prayers for her but did not dare invade her watery kingdom. They threw prayers and hurled supplications at her. She disappeared and emerged. She rose and grabbed the water in her hands, her tears mingling with the sand and the threads of the short, twisted limp hat. She pulled at it and released it, and her muscles relaxed a little. Grandmother's body, arms and legs, carried on. The men and women came. They entered the embrace of the water and embraced her and lifted her. The water wet their faces and streamed down like drops condensing on a cold jug. She was livid and pale, and frightened, but still radiant. She coughed and wiped her face with the black cloth. She kissed and smelled it, and slept with it in her hand. In the morning she took it up to the roof of the house and laid it in the sunlight, spreading it out, drying both sides, combing its velvety fabric clean of dirt, mud, and sand; then she shook it. She kept on talking to it and calling to it. She raised it on to the empty coffin herself and they carried it through the narrow, filthy lanes of Ali al-Gharbi. They wound through the streets near the municipality building. She led the way, walking before them, with Jamil to her right and Farida, tiny, bored, and troubled, to her left. They all walked, the widows of the six men, the women of the neighbourhood, their sons, their daughters, and their old men.

Six coffins, one with the hat like a banner, swaying and shaking, stopping when the crowd stopped before the front door of each dead man's house. The funeral procession walked to the cemetery. They dug up the earth and its cold, damp smell rose. Suddenly, before the burial she removed the hat and held fast to it: "No, I'm taking this to Baghdad, and I'll bury it there." She made a glass box for it and put the black hat in it. She closed the box, lifted it with her hands, and placed it before her. She greeted it before going to bed, touched it first thing in the morning, and moaned at it at night.

The day they left for the capital, she carried the box next to her chest. She distributed all the things to the people of that area. They rode in a taxi. Her children were silent, and in her hands she held the missing man's only remaining words.

My grandmother's family had a large crypt. It was clean and spacious, with several levels, located near the entrance to the cemetery. It was surrounded by a thicket of oleander trees, dusty from hanging in front of the large opening on to the main street. The family tree was hung at the entrance, all the boughs and branches, going down to the earliest roots. My grandmother's forebears came from the Hejaz, and my grandfather's from central Iraq. This crypt belonged to my grandmother's tribe's kinsfolk, and no strangers were to be buried here.

My grandmother chose a spot behind the crypt. She called for Muhammad the builder, and he opened a new grave, surrounded by a cheap metal barrier painted white. Trees of a sort I was not familiar with were planted flanking it; they had short but plentiful branches, and a mysterious, penetrating smell on hot summer nights, a smell like laughter and tears. The hat was buried there, and every year the grave was repainted, the stones and mud were rearranged, the boughs were trimmed, and the plants were watered. She stood, tall, pale-skinned and sobbing, blotting him out with supplications and sending him prayers.

Farida stood this whole time, looking waxen and rigid, as if venturing into a trap. She did not lament or cry, sob or wail. Her lips moved slowly, and her face showed the shock of terrible sorrow. Grandfather had loved her so much, Mr Jamil had isolated her for a long time. That Munir had vanished and not come back anywhere near her. She stayed inside the house for three days. She struck her head and her voice sounded with all its hoarseness and power, echoing through the rooms, even reaching other houses.

All the voices lamented and fell silent, and Farida never grew weary. Iqbal was before her and Munir and Jamil behind her. The neighbours; the women; the rumours whispered from mouth to mouth:

"Munir is never coming back."

"They say he knew a week before the wedding the news of the deceased."

"Why didn't he say so?"

"By God, we don't know."

"Every day he and Abu Iman get drunk. They set up a table in the bar then Munir would bid Abu Iman farewell on the street and disappear."

"No, Abu Iman says he'll get married when he's past forty."

"They say Umm Jamil has fallen apart completely, with Iqbal and now with Farida. Abu Adil has deserted them – this new woman and the children have taken him."

Grandmother went down into the crypt. Farida hesitated, then followed her down. Adil touched the dirt under the oleander, pulled off some leaves, and threw them on the ground. Behind the window, I looked at my grandmother as she offered her prayers to the dead.

It was the first holy day with Iqbal absent. We waited at the door to the crypt. The women of the neighbourhood, aunts and sisters, their faces slack, their complexions changed, eyelids dewy, eyes meek, their cloaks dusty; but they stood on tiptoe. Their hands clutched the arms of their little sons and daughters. The poor *sayyids* waited for their holy day donations holding out their hands and murmuring prayers as they waited for the sacrificial meat of the feast. The sun did not keep Adil waiting; it rose swiftly and was hot. The sky dispersed its clouds. Umm Suturi stood before us, a palm leaf tray in her hand:

"Here is brick oven bread, eggs, and boiled potatoes. Eat them in the train, and remember us when you pray to the Lord of Martyrs. May the sorrow be lifted from all our souls. God be with you."

I carried the tray on my shoulder. We took the bus to the station at Bab al-Muazzam. The trains stood there, rusty and peeling. People were smiling. The pedlars shouted. The boys and girls boarding before us wore colourful holiday clothes. There were brief goodbyes and stifled weeping, and then we went up the steps. We sat facing one another.

Adil and Grandmother, and Farida beside me. The train filled up with soldiers and luggage and the smell of food. My aunt pulled her bags in. We had our first bite, and the sound of the train as it began to move relaxed my bones and made everything I was experiencing seem small. From the wide iron window which was spattered with grease and the remnants of dried snot, I saw creatures – creatures whose double faces, and faces stripped of features, passed before me. I waved to them with a morsel of bread in my hand, knowing that I would never see them again.

12

୧৩৩৩৩

I listened to the noise and yelling, the crying of children, men blowing their noses, and the shouts of the mothers in the corner of the compartment, passing out objects and snacks. They sat on their old suitcases, which were lashed shut with thick, frayed ropes. The women's heads were covered with black bands, from which twisted threads hung down, new and clean, reaching as far as their eyelids. Some pushed us inside and sat down, crowding near Farida.

Adil and I looked at one another. No face looked like Mahmoud's. No girl limped like Firdous. No odour from anyone's mouth was like my mother's. My grandmother drew out her black prayer beads and began to tell them, paying no attention to her surroundings. My aunt picked at some morsels of food and put the rest in a bag, but Grandmother did not touch even a crust of bread. Her cloak was wound all round her body. She watched Farida, and said, in a soft but firm voice: "Wrap your cloak round your body well." The men's and women's eyes stripped my aunt of her clothing. I looked round at everything about me. The man with the headropes looked like Haj Aziz, but his face was older and less bright. I watched the man sitting far away rolling tobacco in paper, moistening his lips, swallowing, and looking at my aunt, lighting his cigarette, sighing, and then raising his voice in an old southern song, in which he was joined by the soldiers heading home on leave. Most of the women looked like Umm Suturi and Umm Aziz.

A voice sounded, alone, from a woman we could not see in all the confusion: "Whoever has not made the pilgrimage to Lord Hussein has wasted his life!"

Laughter, shouting, and singing. The men's cloaks, and new trousers and jackets. Men's trousers, wrinkled, ironed, old, coloured, long enough to touch the floor, short enough to see holes in socks. We smelled the stink of feet and the odour of sweaty armpits. The women shouted with joy and trilled as they recited the names of Ali Ibn-Abi Talib and his children.

Food appeared: skewers of kebab, grilled goat's testicles, and flat loaves of bread that had become cold and wrinkled. Onions and green tomatoes. The movements of chewing and swallowing in front of me made me join them, and I asked one of them for half a piece of bread and a skewer of kebab. I reached out and took an onion, sat among them and ate. I did not look at my aunt. Everyone was belching.

The boys and girls wore cheap clothes, and their shoes were scruffy. Their socks were uneven – one high, the other low. The girls' ribbons hung down to their chests, and their necks were bare and spotted with grease. I did not know what to wipe my hands on, so I left them as they were and looked at my fingers. I got up and walked back to my seat. Farida was wrapped up, but left part of her chest visible. I looked at my grandmother. She had said before we left that "You will wear an abaya when we get to Karbala." I saw my abaya underneath the containers of food; Umm Suturi had brought it. I saw the men above our heads and around us. The young men were smoking, coughing, and staring. I turned my head towards the window.

My father had come from Karbala the previous year. He placed a quarter dinar in my hand and said reluctantly, "Take Adil and go play on the swings. Hold on to him tightly when he's on it. If anything happens to him, I'll kill you."

Iqbal stood silently at the door of the house. She drew another quarter dinar out of her neckline and buried it in my hand, and pushed us outside without a word.

This was the first holy day I had a new dress. It was yellow, and the waist had a shiny belt of delicate satin, a scrap from the cloth of our new quilt. New yellow ribbons adorned my braids. Umm Suturi had stitched my dress in two hours, and Farida finished sewing the back and the sleeves. I was walking, picking off threads and blowing them into the air. Nuriya had sent Adil his new clothes from Karbala.

Firdous and Mahmoud stood in front of the door to their house, Suturi, Hashim, and Nizar waited in the spacious lot behind our houses. That is where the girls and boys of the neighbourhood celebrated.

We walked round the grimy ice cream carts, whose rusty wheels stopped almost as soon as they got rolling, so the ice cream vendor had to hit them to right them. We stood by them. Inside them were large tins surrounded by crushed ice dyed red, yellow, and brown. Small tan-coloured plates and old spoons. The man sold us some and we ate it. We crunched the ice between our teeth, turning our lips different colours. We reached in for a second tin of cola in their dark-green bottles. We kept the cold in our mouths and went to see blind Umm Aziz, who had enlarged her palm platter and placed coloured lollipops on it, spun sugar attached to thin sharp sticks, all on another palm fibre platter. We stood in front of her and started our game here. We wrapped scallop-edged five-fils coins in glossy silver-coloured paper; we did this well until we had covered the milled edges, so that when she felt each coin she thought it was a dirham. She was fooled, and we took everything on the trays. A few minutes later, all of a sudden, her voice split the air cursing us and our parents. Adil went back and gave her all his money. Mahmoud, Hashim, Nizar, Firdous, Suturi, and I licked the lollipops and threw the sticks on the ground, putting the candy floss in our mouths, eating and not caring. We went to the fried seed seller and bought dried chickpeas, peanuts, and black and red raisins. We munched them and the ink ran on to our fingers from the words on the old notebook pages in which the nuts were wrapped. The sound of whistles began to lead the way. Paper kites of all colours filled the air. Hands pulled the kite strings and tails, which rose and fell like Euphrates birds as the wind blew. The boys and girls counted their fils and pennies and grasped them tightly. The young men of the neighbourhood stood around in new dishdashas and wide leather belts, keeping their money in linen bags between the waist and stomach. They called out to everyone using the swings. Among the lofty palm trees the heavy ropes waited for our small hands. I placed Adil on one swing and gave him a vigorous push: "Hold the rope tightly, Adouli!" I got on another swing: "Mahmoud, push me as hard as you can – don't worry about me."

My feet flew high up into the hot air. I saw the roofs of the houses and the red buses, the laundry lines and the window panes. My braids leaped with me as I pumped myself higher. I saw Firdous, silent and serene. She watched me go up and down. Suturi pushed Adil and I shouted: "Harder, Mahmoud, harder!" The voice of the man holding the rope: "That's five fils' worth." I tottered as I slid from the sky to the ground.

The ground was dirt, pebbles, and broken bricks. We slid along, raising clouds of dust that got into our eyes. Wagons drawn by skinny horses passed before us. The drivers called out, "One ride, ten fils."

We all got in and stretched our legs out, all crowded in on one another. We all had whistles and brightly coloured paper pinwheels that spun in our hands when we blew on them. Our voices rose in song: "We miss you sweetheart, God we miss you, It's been a long time since we parted." We applauded and made jokes, and shouted in one voice: "Hey! For God's sake speed it up!" The horses looked like Umm Aziz. The cart took us round. The streets had been recently paved and were crowded with people and automobiles. We rode up the dirt dam and went down Royal Cemetery Street. This was where the first Queen of Iraq was buried, the mother of King Faisal II and the sister of the regent. We stood up in school in the morning and the teacher, Miss Nabila, cried in front of us. We all bowed our heads, and they lowered the flags everywhere for forty days. We cried for the Queen, whose photograph we had never seen, and when we went home we were proud to give our families the news: "Queen Alia is dead."

The sun shone into our ears and eyes. We put our arms around one another's shoulders, and Mahmoud's hand went past Firdous's back and reached mine. I grew hotter; his hand was near my braids. Firdous never opened her mouth or closed her eyes. She was stubborn on the inside and shy on the outside, a little taller than you. Her complexion was wheaten, and a violet green lay deep inside her eyes; they were narrow and bright. Her eyelashes were thick but short, and her teeth were widely spaced, with a layer of plaque. Her lips were dry, as if always parched with thirst. When she spoke, she panted, and when she quarrelled, her voice was a shrill shriek. Her jerky breathing crackled. She charmingly mispronounced the r-sound in the back of her throat. When she laughed, she laid her palm over her mouth. When she walked, she drew her left leg back and heaved it forward. Her pelvis had been malformed from birth. She did not play out in the street until she was seven. They called her Firdous the Lame.

The day they moved to your neighbourhood you stood in front of her. You looked into each other's eyes. She was prettier than you. Her skin was tender, and she was plump – and quiet. At first neither of you spoke. She held an old, small, ugly, frightening rag doll in her hand; around it were the remnants of scraps of coloured cloth, charcoal, chalk, string, scissors, and pens. She would draw and sew, smudging and re-drawing the lines of the face with the charcoal, changing the angle of the nose. She held a pen and moved quickly across the cloth. She put earrings on the ears, made some of the eyes blind, distorted some of the faces, carved and cut the cloth of the rag dolls. She made the faces look insolent, like monkeys, like beasts, recalling all the animals in her books, the gardens, and streets. Eyebrows disappeared, eyes danced, teeth broke, and blood flowed on to the rags. Hers was a strange toy, one-legged, or with both legs cut off.

You stood, watching, not getting tired, and she did not look at you: "Sit down. Why are you standing up?"

"Why not come out so we can play in the street? I don't like playing indoors."

She did not reply. Anything she did not like, she did not reply to. Suddenly she opened up the doll's mouth as wide as possible, pulled off one leg, and threw it to the floor. "Look, it's Firdous the Lame, and this is Huda the Shameless."

"Fine, fine. Come here on the steps, we won't go far."

"But stay with me."

I stayed with her. At first she did not believe it. She did not hate anyone for walking on ahead of her, but she confided in herself, and in everyone around her, that she was Firdous, who never waited for anyone to take her by the hand and walk with her. The days and hours passed but the only thing she worked at was her leg. She lifted her dress in front of you so you could see the thigh with the old flesh. She always listened closely for the voice of her small, delicate cells: "Look, everything's quiet now, but as soon as I start walking, it's something else again."

Firdous was something else again. She was best in the class at school. She was quiet, reasonable, and clever, as immersed in silence as if constantly drunk. She kept her dignity and never relinquished it in front of me, either. In the street, no one ever again dared to call her Firdous the Lame. She abandoned herself to me and I led her, hugged her, and she bore her reputation and mine too. She did not like to make acquaintances or to meet new people. Her curiosity went down to her limbs and stayed there.

Everyone recognized her steps when she came to the house. We went from one class to the next, from secret to secret, and changed. I lifted my arm so she could see the downy hair of my armpit. She looked timidly, then began to count the number of hairs. We stood together, measuring our heights, arms, plumpness or skinniness. The layers of sound and passage of secrets from mouth to mouth. The murmur of breasts, the chastity of speech about the absent children of the neighbourhood. She had time for dreams, and assessed the boys of our street: "Mahmoud is yours. Adil is shy and sweet. Cross-eyed Hashim makes us laugh. Suturi the bird boy is a devil like you. And *he* is just for me."

He was Nizar, one year younger than she, but taller, uglier, cleverer, and quieter than everyone else. She had not made approaches to him; she did not know how to reveal the secret. She stirred up her imagination with it at first, then hung around him, not wanting to deny anything. She was jealous of everything and anything, in a way that we did not know how to prepare for. *He* was hers alone. She talked to herself about him every day, in front of me and when I was not with her. Pretending to be talking about herself, not about him. She tried to quantify his soul syllables through the number of letters in his name; she multiplied them by the number of letters in her name, then added the remainder, and Nizar appeared before her like a treasure. She always said, "Him." She was terribly benevolent towards him, always saying:

"He's sensible. I don't like good-looking boys. It's almost that good looks are scary. We're a lot alike." I said nothing, and she went on: "Sometimes I wake up at night and look at Mahmoud while he's sleeping. Mahmoud is nice-looking, I know, but I don't see him that way. Everyone looks nice when they're sleeping. I only like the quiet ones. Nizar is quiet – as if only I understand him. As if he talks just for me, and is quiet for me. Sometimes Mahmoud is like Nizar, and sometimes he talks a lot."

"When he talks, what does he say?"

She understood you immediately and replied, "He doesn't say anything. When your name comes up he goes quiet. fine – silence is better than saying the wrong thing."

"I don't understand."

"My mother, for example, doesn't like you. She says, 'By God, if Huda were my daughter I would lock her in the house and not even let her see the street.' "

"And your father?"

"He says, 'If God had created Huda in Adil's place, it would have been better.' "

"And you?"

"I'm not ashamed of being lame in front of you."

This was the first time she used the word that made up her title in front of me. I never saw her tears, but I cried in front of her, in front of Mahmoud, and everybody. She left me as I was, crying until I thought my hair and lashes would fall out. I sweated and trembled but she did not come to me. She did not touch me or dry my tears or wipe my nose. She did not think about me, or laugh when I laughed. She was sceptical and high minded. She stretched her leg out before me, spread her dress out, and covered her knee, moving as if she were feverish. Strangely, she looked like our class teacher, Miss Qadriya. She looked at me and at the other girls as if seeing them for the first time. She laced her fingers together on her chest and did not move. She looked at me as if I were one of her dolls. I did not envy her or hate her, or look at her, or forget her. Before me, behind me, her looks, her breaths, her appearance. Firdous came to me, taking the first step with only her left foot. She showed it to me. She got up, moved, and took her first step with her arm against the wall. When she reached the steps she stood there. She stretched her neck into the lane and took a calm look all round. The neighbours' houses; muddy streams; the pregnant house cats. Alley cats from other streets. The neighbourhood men walking by. Housewives sitting on their stone steps. The street in front of us was important and confusing, crowded with bodies and ideas, radiating fear and illusions. She stood there, leaning her pelvis against the doorway as I stood beside her, hand in hand. Our hands were touching. She did not clasp my hand, but let me clasp hers. She never hurried for anything or trembled before anything. I never saw her afraid. It was as if she had bent the tree of fear under her arm and torn off its leaves, eaten its branches, still waiting in its shade for something greater than fear.

Firdous. This was the first holy day where I had not taken her in my arms and kissed her. She gave me a present of money, saying, "Spend it – buy everything you want for yourself. Only don't make me play on the swings."

"What if we play on them together? Would you do that?"

She did not answer. The five boys pushed us and she clutched the rope in one hand, with her other arm around my waist. She never batted an eyelid. She did not falter, but her face became livid and then went pale. We swung up high, and I only looked at her. She looked only up at the sky. We swung down and flew up. I shouted and sang as I watched her feet in the air in front of me. I knocked my shoes against hers and pinched her leg. She did not hit me or push harder against me. She did not frown at me or see anyone before her. Her eyes were fixed. Her lips were dry, and her voice could not sound, no matter what. When the swing swayed, the boys crowded around us and we slid off. Everyone laughed, everyone but she. Nizar came near her, and they looked into each other's eyes calmly. She took a cold bottle of drink from his hand and said, "Thank you, Nizar." She walked alone ahead of us. She stopped but did not turn around. I reached her, panting. She said, as faintly as a voice coming from a well, "I wish Nizar had agreed to ride the swing with us."

This train looked like the swing. We stopped for a long time, then walked along slowly. They boarded and disembarked at the stations.

The voices of the pedlars selling cigarettes, chewing gum, and cold drinks. The stations were ruins, workers' rooms demolished in the middle. The rail employees in their dark blue clothing boarded the train, checked tickets, coughed, and looked at my aunt. I stared hard at them. I got up several times, and Farida pulled me roughly and pushed me down by the window. Adil had not run out of patience as I had. He was tired and fell asleep on my grandmother's lap. I watched him and thought him to be more beautiful than the birds flying before me, the low houses painted bright colours, yellow, black, and a dirty shade of turmeric. The trees stood alone, naked, and dry, not moving as we passed them. Shops and garages were halfway open. Old, overturned automobiles, bicycles boys had dragged through the streams. Iraqi flags, limp in the heat and warm air, hung over police stations and official offices.

My grandmother had still not smoked or had anything to eat. I said to her:

"Grandma, I'll get you a snack. You haven't eaten anything since last night."

"We'll eat kebabs in Karbala, at the shrine. They call them Karbala kebabs."

"What about my father?"

"What about him?" asked Farida crossly.

"Will we go and see him?"

Grandmother stroked Adil's hair, not looking at us.

"We'll bring you and Adil to him, and we'll go to the holy shrine."

"And after?"

"And after?" she said severely. "We are not going to his house. If he wants to see us, let him come to the shrine."

"But we – "

"What about you? If he takes you to his house, go. Your new brothers are there. One-eyed Nuriya is there. By God, she killed Iqbal." My grandmother's voice was clear and decisive: "God took Iqbal. Don't listen to this talk. Give him our regards. Kiss his hand, and tell him God will bless him if he does honest work. I'm longing to see him and hear his voice. I want him to come to the house. I'd accept it if he got upset, or if he got drunk and the men carried him to the house. I'd accept it if he beat you. Tell him, 'Your mother wishes you health and happiness and prosperity.'"

She lowered her voice. She removed her spectacles and wiped away her tears. Adil shifted in her lap. He hugged her and sighed deeply on her breast.

"Come here. Where are you going?"

"Let her walk about a little."

Adil came after me and walked behind me. "Stop a little."

We jumped over the luggage. Everyone's eyes were on us. Their faces inspected us. We stood before the window, finding a place among the young men and girls. We stuck our heads out of the window; the hot air blinded us as we staggered and bumped into the others gathered at the window. I saw numbers of flies settling on the glass and on the nostrils of the people around us. I shooed them away but they came back.

"It's one-eyed Nuriya who killed Iqbal?"

"Are we really going to my father's house?"

"If you want to go, go."

"And you?"

"No."

"But if he takes us, what will we tell him?"

"Grandmother wants to see him, even if just at a distance. We'll tell him that."

"He might get cross and not come."

"He might come with us."

Iqbal cut my father in two.

The train stopped here, at Sakkat al-Hindiya, for a long time. It was the first time we had visited Karbala and the first time we'd ridden in a train. The call to noon prayer, the figures spreading out rugs and carpets on the floor in front of us, facing Mecca.

"Grandma, we'll get off for a little while here."

Farida replied, in her gruff voice, "Stay where we can see you. Don't go far."

The air burned us as if it were coming from a furnace. The tall trees around us surrounded the rest stop at Hindiya. Naked youths swam in the deep brooks, and women dangled their feet in the muddy water. Some of them were bent over, washing and rinsing clothes and squeezing them dry. They scrubbed dishes and metal pots with mud, washed them off, and turned them over on the ground.

Sheep, cows, and goats wandered before us, drinking from the other end of the brook, making their sounds and eating the green-yellow grass. The sound of love songs came from the other side of the brook, interrupted by loud cursing. Adil did not move; he was standing underneath the window of the train. He was watching me dipping my hand into the brook, washing my face and looking at the women, who looked back at me and laughed together.

Again the train released its sound. My hair was matted with sweat and my clothes stuck to my skin. I smelled my armpit. My aunt had a disgusting smell, like burning excrement. I would sit far away from her. "We'll be in Karbala shortly."

Every time I heard my grandmother's voice I thought I was hearing it for the first time. Adil left me space next to him. The toilets on the train were far away. My aunt said, "They're all filthy and full of diseases."

The sky looked like my father's face. We all rocked forward and were pitched on top of one another. We were at Karbala Station.

My aunt attacked me with a stinging voice before I disembarked: "Where are you going? Come back. I swear, if your father saw you running round like this he'd kill you in front of everyone. Take this cloak and wrap yourself in it the way you're supposed to."

"Hold Adil's hand tightly."

"If either of you gets lost, say, 'We are the children of Officer Jamil al-Maarouf.' "

I stumbled and fell, and Adil laughed at me. After a few minutes of walking I began to scratch my head. Every moment I put my cloak in order, it immediately tumbled from my head. My grandmother's voice was lost amid the clamour of all the automobiles and the holiday noise. There were throngs of innumerable people. A woman who looked like a black cloud moved in front of us on the ground, so all we could see were some of the colours above people's heads, the children's white and blue dishdashas as they rode on their mothers' shoulders.

My grandmother and aunt dropped their veils over their faces. Now we could only distinguish them by their voices. The men in front of the shops wore white clothes and undershirts, and all their wares were spread out: rugs, carpets, fabrics of every colour, gold, swords that glinted whenever the sun caught them, fruits, vegetables, watermelon slices set out in rows on large platters, and glasses of cold laban. There were bookshops and shelves of thick, dark-green books whose titles were written in gold. There were pictures of Imam Ali, behind which forked swords shone. I forgot to draw the cloak around me and one of the women smacked me on the chest and kept walking. We stopped behind them and all got into a horse-drawn carriage, then sat opposite them. My grandmother said, "Take us to Karbala Prison."

"Yes, today is the holiday visit. Who have you got over there?"

"My father," said Adil.

"God willing, he'll be safely released."

Grandmother, who was praying, replied, "No, he works there."

"Hmm."

He lashed the horses vigorously, and they led the carriage at a run through Karbala's paved lanes, high and bare, filthy and hot. We went a long way and emerged outside the city, where the air was dusty but the sky revealed. There were no plants, no trees, no houses or garages, no cars, no donkeys. The soil was as white as lime, and the fine, delicate dust settled on us. The carriage crushed the pebbles as it ran over them on the long dirt road.

"By God, I'm only taking you there for the children's sake. No one goes there at this hour."

"We'll drop the children off and go back to the shrine with you," said my grandmother.

"This is the prison. We're here."

Adil's voice: "I'm afraid, Grandma."

Grandmother took him by the head, hugged and kissed him, and I pulled at him. We got off. The cloak fell to the ground. I picked it up, brushed it clean, and put it on my head, the tray in my other hand.

"Listen, if you don't come back soon we'll leave."

"But if – "

"You'll come with us to the shrine. We'll spend the night there."

We took our first step on this ground. We could see the faraway building; it looked like an upside-down lorry and had a high wall the colour of used iodine. All I could see behind it was the sky, with creatures dispersed around it, whose cloaks shone when the sun caught them. Children turned their heads toward the gates which were higher than the gates of our mosque, wide and intimidating, with iron plates in the middle and on the sides and round iron rings from the top to the bottom. The children played with them, poking their fingers inside and pushing their bodies against them. There was a huge hole in the middle in which I saw a key that did not move.

Two jeeps were parked close together in front of the gate. Women were leaning on them and some children were asleep inside. The doors and roofs were open. There was a smell of burning rubbish whenever the wind blew, and the rancid smell spread.

Adil walked in silence, playing with the pebbles and kicking them away. All eyes were on us, and did not leave us. I stood at the gate and placed the tray on the ground, letting the cloak slip down to my shoulders. I looked around me. One of the women asked in a low voice, "Do you have a watch?"

"No."

"Visiting hours start at three."

I knocked at the door and the children laughed at me and crowded around me. I looked at the movement of my palms, as if they were the wings of a fly on the verge of death. I lowered my head to the big hole and shouted, "Mister, we are the children of Officer Jamil."

The children fell silent, and the women turned away from us. A few minutes later the door creaked open sharply and the face of a police officer appeared before us.

Everyone moved towards me in a wave, standing and surrounding us in a circle. They grabbed us by the shoulders and pushed us away, and the man wheeled around, searching for us among the throng. My cloak swept the ground, and I grasped the tray and Adil's hand. The man drove the people away and walked on, holding our hands and pushing us ahead of him. He turned to them irritably:

"How many times have we said visiting hours start at three?"

Before going in I looked back to see the carriage. My grandmother's head looked like an eagle's. I waved to her. We then entered, and the gate closed behind us.

13

༼◉◉༽

"So you are Adil."

He did not reply.

"And you are Huda."

"And you?"

"Jasim. Sergeant Jasim."

"Is my father here?"

"He's here, but he's doing inspection."

"So he's here?"

"Yes."

I calmed down a little when I heard what he said. I thought of cross-eyed Hashim. When he grew up he would look like him. He wanted to help me; he tried to take the tray, but I refused, moving it from one hand to the other. The *abaya* was in my way – I stumbled in it; it twisted round and opened up, revealing my thin body. Before it fell down, Adil helped me lift it and took the tray. When I wore it, I looked like Firdous's ugly and comical rag doll. Pulling the *abaya* from the ground I wrapped it round and looked down at my shoes.

Adil walked. I could not tell whether his feet were pulling him or he was dragging them. My knees knocked together and my fingers trembled as I clutched the cloak; I tried to swallow but could not. If only I had tried my cloak on before. Oh, we had done that before, Firdous and I, we laughed for a long time, calling out in voices like the voices of my aunt and her mother, and we quietened down before anyone came in.

The sergeant walked quickly, then stopped to let us catch up, then resumed walking ahead of us.

It was a long, open path, paved only with old footprints, and shiny pebbles both small and large. The soil was red and rippled, covering and uncovering itself as the wind opened up hollows and then filled them, forming mounds and throwing them up into our faces. The edge of the cloak flopped over my face and I nearly fell, but Jasim grabbed me from one side and Adil grabbed me on the other and we stopped. I closed my mouth, pursed my lips and smelled the sand that had got into my hair and between my toes.

"He will be delighted to see you two."

Calcite air, yellowish, white, red. The wall was behind us, and rooms I could not count before us, distant and small like sand in a swollen eye.

I did not distinguish the colour; I thought I would ask Jasim about it but I kept quiet. I said to myself that it was perhaps the colour of cooked olives. Sweat had begun to trickle down the top of my neck to the top of my spine, and I felt it moving down my back. It ran and I did not know how to stop it. Now my scalp began to itch as well. I reached up to scratch it and then I did not want to stop. Yesterday I had washed in the bathtub at home; it had been months since we had gone to the market bath. When I came out of the bath Firdous stood in front of me and said, "When you grow up you're going to be tall, and you'll look beautiful in a cloak." I was still short, and this cover made me smother and stumble constantly. My skirt was black, as was my blouse; I had borrowed it from Firdous the night after the death. My waist was wet with water. I touched my middle and let go of one side of the cloak; it dropped down, and grains of sand and sweat wiped off my hand.

There was not a single tree in this whole expanse. I did not know what time it was; my watch had not worked for months. My father had brought it for me when I entered the third grade in elementary school. I did not mind all this walking, but my thirst, and the tribulation of my bladder! How was it that Adil had not yet asked to urinate? How could I ask? He was so patient and reserved, and could piss on himself if he saw my father in front of him.

Now the inner gate of the prison was in front of us. It was wide, and high as well. We entered the way cats enter the gate of a mosque; first voices emerged, then the men appeared, their moustaches, their big black boots notched with nails, the smell of their armpits stirred me up as I stared at them. They turned their heads. One of them was clean shaven, and his skin was discoloured, as if the sun had never shone on him for a single second.

We walked down a long, dark corridor. Men walked in and out, turning around and looking. The floor was of old yellow broken brick scrubbed very clean, and reeking of disinfectants, yet flies buzzed all round us, heedless of cleanliness and unintimidated by the police. They buzzed round the men's noses and bare heads. No one shooed them away or killed them as we did in our house.

"Please come in. This is his room."

We stood in the middle. Adil walked round a little and placed the tray in the middle of the room and sat at our father's table. I looked all round me, turning and glancing about. I went to the mirror but saw only the top part of my head. I stood on tiptoe, stumbled, and fell, me and the cloak. I looked at the floor, stained with this piece of diaphanous silken cloth. I gathered it up in my hands and threw it on the only bed in the room.

The floor was of tile that had lost its colour, and become mud-coloured. A thick-sided rectangular metal table with three lockable drawers on the right hand side, and an orderly stack of papers on top. My father loved order. Envelopes and folders were piled neatly. There was an empty waterglass with a sandy residue at the bottom. An old ashtray with a recently stubbed out cigarette, which I emptied in the rusty wastebasket and returned to the table. A black telephone whose surface was smudged and cracked where the numbers were; it was antique, and bore the royal emblem. A wooden chair with a wide back and square pillow with some of the dirty brown cotton stuffing spilling out of it. Adil rested his head on it. On the high wall above the chair hung a picture of the King of Iraq and the Regent on the throne, both in brilliant white clothing. The picture frame was old and silver, and slightly dusty, even as I viewed it from a distance. I approached and wiped it with my arm to see: the King of Iraq was still, and the Regent was showing his even teeth.

Adil tossed his head back. I walked over to the only window, which was also painted in a dark colour, and had a cheap wrinkled curtain. I stood there with the smell of wild thorns passing over me – they were massed like a second, outer window. There were iron bars over the window. A narrow black water hose passed its voice over the tops of the thorns and through their branches, moistening the hot air, soaking up the dust and dirt. It gave off a light, secretive smell of cold that entered my ribs, dried my sweat, and rose to my head. There was a ceiling fan whose sound, as it turned, was like Firdous's voice when she talked. On the other side was a very low iron bed covered with an earth-coloured sheet, and to the side my father's blue dishdasha and cloak, and below them his big leather shoes.

A dark-coloured but clean sink was in front of us, with shaving implements on the cheap metal shelf over it. On the wall, a faded towel hung from a big nail.

I lifted the tray and put it in a corner, and sat on the bed.

At once I felt sleepy. The shade was lovely and the air was heavy, and there was no sound from outside. I removed my shoes and placed my bare feet on the floor, and saw my footprint there.

Adil and I did not speak or watch the door. If we had been left there we would immediately have fallen asleep. Would our father be angry with us if he saw us here? If he did get angry, would he hit us in front of the police? No one had hit me for long months. They said I had grown, and it was wrong to hit a girl who had come of age.

Puberty: the unknown door had opened before me, and I saw drops of blood on wide, unbleached clothing. I was not frightened. I had seen your blood flowing from your nose, legs, and mouth. That was my first blood, the exclusive possession of Officer Jamil. This blood would be yours alone. I took off the clothes and looked at it for a long time. My grandmother and Aunt Widad had trained me, and it was concluded in secret. They said: "When you become of age, you should fear men, all men. You can be a mother or a goddess." I was terrified:

my mother was dead and I did not know anything about goddesses. It was not the blood that frightened me, but masters' complexions: Jamil, Munir, Abu Iman, and . . . they all came out of the secret suffocating rooms and began to spray you with hoses of fire. You inscribed your clothes with your slender, delicate fingers, locked the door on yourself, and left the blood before you. You looked at it as if he were a new brother of yours. This was your blood, and the first time it came out you did not strike or scream.

Sergeant Jasim came in carrying a round tray with two glasses of laban. He placed it on the table, and I went to him. "Sergeant Jasim, Adil wants to wash his face, and – "

"The washrooms are at the end of the corridor, on the right."

Adil paid no attention and did not move. His head was hanging back, as if he were dead.

There were droplets of cold water on the sides of the glasses. The thick rich froth got on my lips as I drank. Adil drank but said nothing. My father's voice sounded behind me; the glass trembled between my mouth and my hand, dripping on my clothes as I set it down on the table and turned to him. He went first to Adil, and round the table he took us in his arms and hugged us tightly. Had my father grown shorter? Or had I grown taller?

Adil began to cry and I did not know what to do. Not one tear would fall, not one word would come, and he was more perplexed than we were.

Adil's voice was the first to crack: "Papa, my mother is dead."

Sergeant Jasim's voice, as he saluted my father. I heard the sound of his legs as they rubbed together and he raised his arm: "At your service, sir."

We clung to him, both turned toward the sergeant. He looked at us and lifted Adil to his chest, and carried him over to the bed. I walked behind them.

"Have you eaten?"

No one answered.

"Go and bring twenty skewers of kebab from the town."

He left Adil, took out half a dinar, and went to the table, took the glasses in his hands and came toward us: "Drink the laban now."

He did not look into my face or Adil's, but reached his hands out to the sink, turned on the tap, washed and dried his face, and took off his jacket. There were splotches of sweat under his armpits, and on his stomach and back.

He sat beside Adil and stretched out his legs. I slid down to the floor in front of him and looked at his feet. Instead of his boots he wore ordinary shoes, which I unlaced and pulled off, but when I began to pull his socks off, he pulled them back up and said, "Thank you, little Huda, we're going out shortly."

He ruffled Adil's hair, stretched him out on his lap, and petted his face. They looked at one another. He lifted his face to him as I stood before them: "Have you taken your school certificates yet or not?"

"Papa, Adil passed, and I – "

He took me by my hand and pulled me to his side, and put his arms round me. My tears streamed down, and my father cried as well. He took his hands away from us and raised them to his head, covered his face, and the sound of his sobbing grew louder and hung in the room's hot air.

This was a face I had never seen before, and all the moments and old images came near me. His haggard face, the delicate strands of grey more plentiful in his hair, the despotic appearance that aroused our aversion and hatred. These were his tears; he had not borrowed them from someone else, and he was not covering them with a handkerchief. He did not display them, and we could only see them up close. If only Iqbal knew; if only Wafiqa knew; if only the whole neighbourhood knew, that Officer Jamil was covering us with his wailing and his charm. We were crowning him now as father over our small heads, and he was sealing them with white wax and accompanying us as we crossed the road. No pistol with which to humiliate, no whip scourged our skin. Jamil had stopped crying, and we stopped studying his head; we held him by his arms and took him by his sides, and turned to him. We squirmed into his embrace and he hugged and kissed us on the neck and hair, smelled our ears and mouths, and a tear fell from his eye on to our hands. We cried as if Iqbal were there with us all, released from prison and free with us. He stood us up in front of him and looked into our faces, never taking his eyes off us. He dared, he dared us, and got to know us; all that was before us was tears and sorrow and fright.

126

My father changed his clothes; he surprised us and we saw him change. He kept us waiting, and joined us halfway.

My father.

We grabbed him and shook him and stood together and pulled him to the sink. He blew his nose, washed and groaned. We were behind him. I grabbed Adil, wiped my face with my hand, put the cloak on my head, and we went to the washrooms.

We went back and found him stretched out on the bed, his face washed clean, his eyes bloody, his mouth about to speak. We stood at his head, Adil stayed near him and I wandered round alone. I picked up the tray and went to the table, took out the bread, peeled the eggs and potatoes: "Papa, will you eat with us?"

"I'll wait for the kebab."

"Grandmother and Auntie will eat kebab at the holy shrine."

He spoke in a very soft voice: "How are they?"

Adil stood in front of him: "Papa, why don't you visit us like before?" I quickly added, "They send you their greetings. Grandmother prays for you all the time when she says her prayers. She raises her head and says, 'Soften Jamouli's heart.' She wants to see you. She said, 'I'd accept him coming even if he got upset and beat you two,' Papa. They are at the holy shrine."

Sergeant Jasim came in. He did not see my father in front of him and did not know whom to salute. My father stirred on the bed and then stood up.

The smell of kebab, onion, and chopped celery. I opened the bag, and a light vapour emerged through my fingers. A layer of fat was stuck to the bottom of the bread. Red sumac was sprinkled on the skewers of kebab and wilted sprigs of mint. There were sharp Karbala-style pickled vegetables, cooked in vinegar with hot peppers, cucumbers, rose-hued boiled turnips, and tomatoes. Sergeant Jasim returned with a container of laban and clean glasses. We three ate. It was the first time we had eaten together. My father broke up the bread and put the kebab in the middle and pushed it toward us. Adil's voice: "I've had enough, praise God."

Sergeant Jasim went to the middle of the room. Whenever he saluted I wanted to laugh. His moustache was luxuriant, his complexion was yellowish, his cheeks clean-shaven, the hair of his head was frizzy and he had one green stripe on the shoulder of his jacket. He was short and stocky, and his teeth were white: "Sir, we'll open the gate at three-thirty."

"Leave solitary until I come."

He went and sat on the bed, took his shoes and put on his jacket, put the *sidara* on his head, and straightened it as he stood in front of the mirror. He washed and dried his hands.

"Papa, we'll go with you."

Adil said, "To the shrine?"

"No, not now."

The voices of men outside, the tramping of their feet, their military gait; the gate opened with the movement of large keys and the rattle of iron chains. Adil went to the window and pulled the curtain aside, and looked out. "Papa," he said sadly, "Do you remember when you told me 'Come and see how I live in Karbala, the dirt and black death'? Papa, I still haven't forgotten that."

Adil turned to us and ran to our father, buried his head in his chest, and we left the room.

I had disappeared inside the cloak, with only my ugly, plague-stricken face showing. Whenever we passed people, they stood up and saluted us.

The police came through the doors and stood in the large court-yard, their rifles on their shoulders and their faces expressionless, their lips thrust out, their uniforms sweaty, the sun beating down directly on to their weapons. There was a sudden flash in front of us as we passed them. They watched us, their eyelashes trembling and eyelids twitching. Their arms were not steady, and the vast courtyard rose as one human wave as they moved and turned. The women shouted. They opened their arms and uttered moans and incoherent words. Their tears flowed down their cheeks. Minutes tumbled by these women, things, and faces.

Friends, relatives, fathers, brothers, uncles, neighbours, spreading their cloaks on the floor, looking into bags, handing out food and cigarettes, sharing water and a little money, weeping and kissing, falling silent, watching, as we plunged into their midst. They touched Adil's head and looked me closely in the face.

The faces of the prisoners, slender figures tall and short, their eyes wandering, cheeks sunken, thick moustaches and slack jaws. Their dishdashas were dirty and their sandals cracked. They all became one colourful, wandering, mad planet.

The men looked like the men of our neighbourhood: Abu Mahmoud, Abu Iman, Abu Hashim, and Haj Aziz. We approached, and the gates which had been olive green were colourless. Here they were before me, I felt them with my hand and looked inside: stone steps, thick brown paper, dug up earth, the high wall. flies flew out into the heat and solitude outside. The smell spread outside, like the heat of the baker's oven in our street, like the mud of the Euphrates in the first months of flood.

My father left me, I leave him. Not one stone over another; not one neighbour near another neighbour.

How could that man laugh? When would he urinate?

I counted: one, two, twenty, one hundred. I did not know how to count the prisoners. Mahmoud knew that I did not like arithmetic, but I was able to count them. They sat on the ground near the children. The women sprawled out on the ground though the sun was no good at this time of day. Some laughed, and diverted me, laughing. Was it lawful to laugh in prison?

My mouth had a bitter taste, my lips were parched, and my tongue was dry. It was tea-time. I turned round: my father and Adil were standing before the locked gate. A group of police officers stood round my father. On the other side was the solitary.

When my father was alone with me on the roof, I was alone with the ants, flies, and fear. The iron bars before me were as silent as our iron roof with glass. There I looked down from high up. I saw everything all at once. My mother, as well, had been freed from prison and wooden talk. Here there was no white or grey glass. The gates were high with small round windows admitting fine dust and flies. No steps took me to the high roof, and no staircase brought me down to the doors that opened one after another. A few visitors waited to the rear. Rifles, police, my father's height and his face, also freed from his prison, coming out one after the other. My father turned round before them, Adil raised his head and looked, holding him by the arm.

My father extended his hand to them, took them by the forearm, and walked with them a little. One, two, seven; they became thirteen. They opened their eyelids a little in the light, moving along with a number of guards among them, though at a gesture of my father's hand the police moved away. He produced a packet of cigarettes, walked, shared some and lit them with a match. His hand touched their fingers, and they took deep breaths, they sat and stretched their legs out on the ground. My father turned to us and stepped back a little. We stood at a distance. The families sat down to rest by one another. One of the sergeants approached us.

"Sir, we have thoroughly searched the trays."

With a nod of his head, the food was distributed to them.

We walked behind him; we turned when he turned and stopped when he stopped. Now he was close to us, now far.

Suddenly he turned to us: "Adouli, go and bring the rest of the kebab."

Adil ran in and returned with the kebab, eggs, potatoes, and bread. My father walked with the tray on his shoulder, the cigarette in his hand, talking with the guards and the circle of men who had been around him. We stood together before those men. They did not lift their heads or lower their arms. They were motionless and silent.

"This is Master Abba's kebab, which my mother has prayed over and sent."

He took the tray, placed it on the ground, and walked away quickly. They lifted their faces to face the sun. For the first time I saw their faces. Their eyes were beautiful, their eyelids were swollen, their eyelashes were dry, their hair was dusty, and their fingers trembled as they held their cigarettes. They coughed. At last their lips opened to show their yellow teeth and white tongues.

"Thank you."

We walked behind my father, who was now far ahead of us and entered the crowd. I saw him, calm and contented but far off. I did not know when his fright ended, or who had buried him with titles. I could remember his first slap and his dreadful lair. His face was moist with sweat, tobacco, and iodine. He was the handsome king on the throne of those who sat before me, my good-looking father who had begotten me and loathed no girl like me. One day Iqbal said, "One night your father came to you when you were in bed bloody and sweaty. You were fresh and new. He was afraid when he first saw you. I thought he would change you to mere bones, and you would make him a policeman constantly standing at your door. He hated it when you cried at night and wailed in daytime. He hated your slow shouting and your rapid breathing. When he slept, he imagined you were hitting him on the head, and he'd wake up wanting to hit you, but I woke up and stopped him. He cried, you cried, and we all cried together. I could never fool him, and neither could you. Every woman fooled him except us. I used to lie down beside you, and he'd lie down far from us. He'd mutter between his teeth as he slept: 'I'll only have boys, I'll never get Huda married. I'll have her dedicate herself only to me. I'll have her never grow up.'"

He turned and we turned with him. He pointed to one of the guards, and looked at his watch. He approached him politely, greeted him, and looked ahead: "Split them up now. The visit is over."

My father gave orders but did not have to listen to them. He dragged his feet sluggishly. Minutes of goodbyes and the sound of kisses. Heads drew away from the circles and stood apart from one another, they turned and walked away, stopped, picked up their children, put on their cloaks, sobbing and praying. Leave-taking clogged the space between their noses and mouths.

The last of the women visitors left the courtyard, a poor, bent-over and tearful old woman who walked and stopped, turning around and ceaselessly praying: "God is good, my son. Yes, you are not the only one in prison."

She reached the main gate and spat on the ground, wiped her mouth on her arm and repeated, "God is good."

We followed her out. My father asked for the driver and the car. Adil clung to me. We turned to look back.

The courtyard was empty. Blowing dust sent crumpled leaves flying round us. The pebbles were not shiny, and the ground was dry. The sun was sinking quickly as we got into the car. My father sat in the front seat, with Adil and I in the back. I wiped away my sweat with my arm, and the cloak slipped down a little. I coughed and sneezed. I raised my arm again to my head and looked ahead. Sergeant Jasim was driving us to the holy shrine.

14

❧◉◐

Farida primped every day, furtively, lest my grandmother know, sum-
moning no one, alone in the cold room on the high roof filled with
gloom and sorrow. The night gave her feelings of hatred and loathing.
Months of days. Hours of bruises and slow, repressed rage. Every day
she fed her beauty with bribes and great blessings, never leaving her
bed of indifference. She wanted the appearance of the first scream: a
man and a woman.

Grandmother taught her the obedience of the blood and the merit
of waiting as a beautiful virgin. The white handkerchief was under the
pillow, but she took it out and looked at it for a long time: "How I
hate this colour."

Her black dress covered her body; we saw her as she fled among us.
She took it off when everyone was asleep and threw it on the floor, trod
on it, and stamped on it, starting to scream, and the words mounted
up before her: "When he comes, he'll find out who Farida is! Where
will he go?" She felt the clothes with her hand, holding each one up as
she stood naked before the mirror: "Even my beautiful skin has turned

black. I hate black. I hate Munir. I swear to God Almighty, you are taboo to me until Judgement Day. You'll see."

Munir had long been absent. She did not love him, but now she wanted him. Grandmother had lost her former authority and her calm. She had lost weight and looked emaciated, purer than before, yet with a resigned face. She bore a heavy load on her shoulders. She emerged from rooms and passed down the hallways, went to the kitchen and ascended to the roof. These two women never spoke or went near one another. She knelt and prayed, opened her arms, made incessant supplications, lamented all alone. She went to the Friday Mosque and stood with the poor people, crept in among them and addressed his spectre: "Why did you seal the contract on the Qu'ran? Why did you build the room and buy all those things? Munir, why don't you come? Farida will die and I'll die after her. Come, may God guide you and deliver us from this tribulation."

The horizon of the neighbourhood took shape. Our neighbours and friends, women and young ladies, relations and acquaintances, waited behind their walls for the spurt of blood.

Farida learnt to talk to herself for hours. She looked at herself in the mirror, took a long knife and began to pass it by her neck.

Ready for the slaughter: "I'm the one who will slaughter you. You'll see."

We waited before her, unafraid, not speaking to her or going near her roof. She wandered from room to room, lighting all the lamps to see all the gaps in the tiles, and leaving the lights on until dawn. Grandmother pushed us roughly when we wanted to go up the stairs. For the first time I saw Wafiqa's gruffness, and I kept silent. In Karbala, she had tied green scraps to the tomb of the Lord of Martyrs, had grasped our heads, Adil's and mine, and made him kneel on the holy ground, saying:

"I prayed to the Commander of the faithful, Imam Ali, for Jamil to have a son, and Adouli was born. Jamil said 'We'll name him Adil.' He loved that name so much: Adil was the first son of his first wife. People began to call Jamil Abu Adil. When you came, I chose your name, Huda – "guidance" – because I said, perhaps God will guide her on the true path. Your father said, 'You choose the girls' names.' Patient, reasonable Adouli, was a gift from our blessed Commander of the Faithful."

This white dove stood to her full height in the holy shrine, and cried at the top of her voice:

"I beseech you, Abu Abdallah, lift this cloud from us all. God is most great. What has happened to us? Fine, Jamouli came, and I saw him before God takes me to His mercy."

When Jamil came, she kissed him. They did not speak, but tears streamed down their faces.

I could not find Grandmother there. Whenever I opened my eyes I saw her telling her beads and watching the sick youths, crippled children, and black-draped women. All I saw were clear eyes looking out from a well of tears.

Their chests were familiar; I wanted to embrace one of them and sleep among their clothes. Their arms were strong; their forearms never wearied of clutching the grillwork windows of the tomb, cradling their children and weeping.

The smell of food cooking in the distant rooms wafted through the air to our nostrils, and I remembered my aunt's delicious cooking. I smacked my lips at the broiled Karbala meatballs and longed for Lord Hussein kebab. The fragrance penetrated the stink of sweat and the steam of anonymous bodies that reminded me of Umm Suturi.

Grandmother prayed for her and for Suturi to have success and divine guidance.

Loud weeping, muffled moans, the stifled supplications sometimes not heard; a chaos of sobbing as if drawn out upon the sinews of rope. Wherever I turned there were voices and more voices. Adil squatted like an angel pursued by devils. Farida said all her prayers and relaxed. She wiped her face and left her body uncovered by the cloak. A soft light radiated from the fabric of her skin. Grandmother turned round and round like Mahmoud's top. They called to her: "Who is Umm Jamil? Her son wants her outside." Jamil was before her. She took him in her arms, a queen in her tears, and kissed him on his forehead and hand. He drew her to him and took her hand, bent to her chest and the middle of her belly, wanting to carry her among the women and children as they smiled and stared at us as hard as they could. Farida walked slowly, diffident and sleepy. He kissed her forehead and did not speak to her.

She let him have her head, kissing his hand and sobbing into his chest. "God rest her soul, sweet Jamil. God willing, you will never see adversity again."

Iqbal was among us like the tomb. Munir was present here, like death over our heads. No one asked about Nuriya and the children. Nor did he invite us to his new house. There was a flurry of hand movements between Grandmother and Jamil; he produced some dinars and put them in her hand.

And . . . the months passed and the days crowded. Grandmother was at the market, Farida was on the roof, and we – Adil and I – cleaned the house.

We began in Grandmother's room, turned over the sheets and bed covers; it was the first time I was the lady of the house. I did not even think of Munir, or imagine Farida being poisoned and dying, or Munir being carried home drunk and dying alone. I started with the windows like my mother and issued orders to Adil.

"Oh, if Mahmoud saw me now!"

I had the keys to all the rooms in my hand. I opened them up and stood before the steps to the roof.

Like a prisoner, Adil stripped his bed and lifted the beds and chairs, rolled up the carpets and dusted underneath them; he looked me in the face, and smiled. He was wearing short trousers. His shoulders were broad. All of a sudden he was growing taller, but his pockets were still filled with raisins and dried apricots. Now he had to go to the barber Sayyid Abd al-Latif for a monthly haircut, where his ever longer and thicker locks of hair were cut off, so he looked like a boy whose beauty had been squandered on those around him.

This Iraqi brother, who was dumb with terror in my father's presence, changed as we crossed the ruined dirt dams. We hoisted our bodies and looked over to the other side of the street, towards the tree-lined corniche road lined with tall houses with spacious gardens full of lofty trees, with constantly sparkling windows complete with iron shutters. We stood for a long time and smelled the penetrating fragrance of citrus blossoms, moss, and roses immersed in the water of the Tigris.

There children played table tennis, badminton, or volley ball in special areas of the park. They went down to the river by the side paths of their houses, they swam and went home, always wearing new sandals, and draping snow-white towels over their slender bodies. The girls in some of these houses took private piano lessons, and big buses came to take them to their private schools run by nuns, or the Frank Ayni Jewish School. They never looked back much, and when they looked ahead their glance was a combination of indifference and annoyance at the rows of boys walking on foot and carrying their books in cheap linen bags. These boys kept dusty sweets in their pockets along with a few coins to be shared with their brothers and friends and they went along singing, laughing and whistling at the buses, houses, indeed the whole world around them. Adil gazed at Khulud and sighed: she was an arrogant girl, as beautiful as a foreign doll. If Adil disappeared for hours I knew where he had gone. He would be there; he knew her name and how many brothers she had. He picked flowers for her and he went down to the banks of the river, walking close by her huge house which practically smelled of money, good meals, and long vacations. He put the flowers on the garden wall, but waited for no one.

He went there every Monday and Thursday, walking slowly. He studied at night and flew his kite in the afternoon, writing her name on the biggest one before launching it into the sky. He saw her as a creature who had come from the sky, who had no time for citizens of the earth.

He stood in front of the bus at her school and watched her board as if she were going into the sky. He went from one class to the other, washed twice a week, and legions of sorrows collected in his eyes.

Suddenly the door opens and closes. We turned around and, and a single utterance choked both our throats: "Uncle Munir!" His bald head, his bright eyes, his face, now even darker. He coughed and cleared his throat. He looked at everything around him as if seeing it for the first time. There were more dark lines under his eyes, and I presumed, the same old talk behind his lips. He did not look at us or follow us. Adil did not hide from him, but kept watching him. I was silent. I held Adil's hand and stood before him, never taking my eyes from him. We exchanged looks. He was uglier than before. My hair was tousled and my dress was hiked up above my knees. I pushed it down. My face was dusty. I cursed this Munir, his father, his grandfather, and his sharp, penetrating voice: "Where is my uncle's wife?"

"She went out."

"She went out? Odd. Where to?"

He did not look up. I took Adil into the bathroom: "Wash, and clean your body well."

I closed the door behind Adil.

"Come here," said Munir.

I went to him. He was wearing a wolf skin. I saw a jackal before me, loathsome and sinister. He sat on the mat in the middle of the house, lit his cigarette, and threw the match on the clean floor. I silently asked God's pardon as I brought him an ashtray.

"Don't you see how tidy the house is? You've come back and brought your orders with you. God must have been angry with us to have sent you here."

He started to laugh, a slow, shameless laugh, then raised his voice: "By God, you've grown up. Now you're giving orders. What – didn't you want me to come back?"

"I didn't care whether you came back or not."

"I came back for your sake, for all your sakes, especially you. You have a sharp tongue, and you're saucy and stubborn. I can raise you."

"You have your wife in the house. Go up to her. Raise your voice with her."

I got out of his way, leaving his wicked voice behind me: "Come here. The house is tidy, and Umm Jamil is not here. Were you waiting for me? Hah?" He laughed gloatingly and resumed, "Hah. Why don't you answer? Where has your grandmother gone?"

"I don't know."

"And her?"

Oh; her.

I vanished without answering him. I looked for a voice to sting him; the Qur'an on the radio gave me faith, and I lifted my voice high, chanting with it, moving before him. I went into one room and came out in another; he coughed but did not speak. He smoked, and lit one cigarette from another. I prepared Adil's clothes; his voice sounded from the bathroom as if scaling a lofty mountain.

"Huda, come."

"What?"

"I'm out of cigarettes."

"Go and buy some yourself."

He rose from where he sat and came near me, stopping me in front of the door to my room. He took me by the hand and twisted my arm, and said, almost inaudibly: "If it had been Mahmoud at the door, you would have gone to buy them. Hah, I know everything about you – now you'll go and buy the cigarettes."

"Ouch. Let go of my arm. I won't buy them. Even if you beat me to death. You go and buy them."

"Don't raise your voice louder than mine. Do you understand? If you want, I'll make it clearer for you." He twisted me and I turned with him to release my hand from between his clasped palms. For the first time we touched one another with such strength. "Fine! Ouch!" I shouted. He released my hand. I looked at him. I wanted to spit in his eyes, darkened by such thick eyelashes that even as he spoke he seemed to be asleep. He reached for his pocket and took out some money which he put in my hand. He left me alone and walked away from me into the house. I was frightened when I saw him in my room: "By God, you're getting religious like my uncle's wife. Now hurry up and go." He looked around him, his hateful face, his eyes like a dog's. I went out, muttering, "You should have bought them before you came."

I went to the gate of the house, opened and closed it behind me, disappeared into the passage and waited.

He removed his jacket and dropped it to the mat, then walked down the hallway that led to the bathroom.

In a flash I mounted the steps and stood panting in front of Farida.

"Auntie, Munir is here."

It was as if the gates of Hell had opened. She got up, her face blazing, her lips dry, her eyes bulging out, pushed me aside and raced down. I was behind her. Her voice was like my father's when he dragged me down the stairs by my braids.

"He came and didn't bother to see me? Munir Effendi! Where is he?"

She turned to me: "Where is he? Where is my fine, gallant cousin? Today is your judgement day, Munir. Come out." She began trembling:

"You little bitch, are you trying to fool me? Only you haven't fooled me. Where is Munir? Tell me, or I'll kill you instead of him."

She had my hair in her hands: "I swear, he was here a minute ago." She turned and saw the jacket and his cigarette butts, and looked all around: "Where is Adil?"

"In the bath."

It was as if Farida's face was spiked with thorns; the whites of her eyes were bloodshot and half her tongue was hanging out as she raced to the bath, opened the door, and thrust her head in. I was behind her. Adil looked stung, holding a basin of water in his hand, his face covered with soapsuds. He turned his head quickly and put the basin down on his thighs: "Did I hear Uncle Munir's voice?"

"He opened the door on me a minute ago."

She was dripping with agony. Her piercing voice hunted him through the rooms of the house. She came out, went into the water closet and stood at the door. She heard the sound of water pouring out of the pitcher, and banged at the door: "Come out, Munir. Come out."

In seconds everything turned to terror. I watched her, not moving or uttering a word. Farida had been preparing for months; what she gathered was scattered by her voice, which grew heavier and more tense; her tragedy could be heard. She pushed against the door and banged on it. Then we heard his heavy, mocking voice:

"Wait a little. You're truly mad."

The door moved, and there was a glimpse of his baldness. She reached out her arm and started with his head and neck. She pulled him out by his tie, then pushed him back inside and followed him in. She pushed his head into the toilet and then pushed him outside, grabbed him around the waist, and they ended up in the long, narrow corridor. Their voices clashed and there was the sound of blows.

"Huda, call the neighbours, and you and Adil come – all of you come and see your aunt's wedding."

He escaped from her arms but she caught him. He looked at me and nearly fell: "Surely your aunt has gone mad."

Farida opened her arms and raised her voice: "Yes, mad. You took a year and I waited. Every day I said, today he'll come. Today he'll open the door and come up. Today will be Munir's day."

This was one of the *sayyids*. My aunt collapsed on him, pulled him, and he slipped away. She grabbed him by his shirt and brought him down to the floor as he kicked about. She gathered her rage and screamed, "Even if I kill you with my own hands I won't be satisfied." I wanted to reach out and beat him myself. She shone and whimpered and bent over as if she were in the market bath. She pulled him to the middle of the house, snatched the pillows and threw them at him, stepped on him and pushed him, got on top of him and sat on him. She raved and called out, got up and sat down: "I'm going to kill you with my own hands!"

I watched, and the man hacked and choked. He twisted his arms and kicked his feet. Farida's chest shuddered. She grasped his thighs tightly, and repeated, between gasps for breath: "Come here! Grab his legs with me!" I did not move. She got up and put the pillow over his face, sat on his chest, opened his legs, seized his leather belt, and began to undo his trouser buttons, then pulled down his trousers, worked up like a lunatic. In a flash she stripped him naked in front of you. You watched the movement of his legs as they kicked and flailed.

Everything was before you now: the hunting rifle and the unicorn.

She spat on him and beat him, shouted and cursed him. She bit him and formed fists, raining blows on all his limbs. "I don't want you to die. Death is too good for you."

She beat him and howled.

"Listen, Munir, I'm going to throw you out. Get out before I kill you."

She turned and exploded, then collapsed a distance away from him, poisoned. She knelt on the floor and rent her clothes from top to bottom, smote her cheeks and tore her hair. She wept and wailed, and suddenly set upon him in his resigned state, His feet were still, his trousers were halfway down, and what was between his legs looked like stale meat. Your aunt's voice had trailed off; she wailed almost inaudibly, crying softly but not rising. She called but no one came to her. She exclaimed "*Allahu akbar*," beat her neck and fell on him again, only her arm threatening. She was sweaty and her hair was matted, the bosom of her dress was open with her breasts exposed, now jiggling outside, now back inside. Adil walked through the house, terrified but silent; he did not stop or see anything. He went into our grandmother's room and buried his face in her bed.

Mr Munir felt himself, pulled up his trousers and pushed the pillows away from his head. She went to him and kicked him in the chest and forehead, stood over his head and spat on him as if she were about to vomit. His bald pate shone with spit and sweat, his harsh, wrinkled forehead withdrawn a little. He closed his eyes, covered with spittle. She pushed him in the stomach and muttered and she pulled me by the arm. We went in to Adil and she turned to lock the door with the key, then sank to the floor, beating her thighs, and tearing her hair. None of us made a sound. I thought of my grandmother as I heard the sound of the outer door opening and closing again.

15

Everyone in our street was stopping by the shop of Hubi the butcher, all stunned by the rumours: "The police have taken Hubi away."

"They say he was circulating anti-government leaflets."

"No, they say he cursed the Regent and Nuri al-Said."

"God help us and our children. They say he was behind the last demonstration, after Nasser nationalized the Suez Canal."

We had watched the demonstration: my grandmother stood in front of the Friday Mosque with the women of the neighbourhood, praying for the young men as they passed before her holding their banners high. "God protect you, my dears, and bring you safely home to your families."

She was bewildered, exclaiming as if she stood in the line of fire. Umm Suturi belted her wool cloak round her waist, stretched, and tightened her black band round her head, trilling. She regulated the water spigots, set up five thick wooden posts and set pots and pails of clean water between them. She filled canvas sacks with loaves of oven-fresh bread. They drank as they passed before her, shouting slogans and munching the fresh bread.

Rasmiyah had prepared a number of emergency supplies: surgical spirit, dressings, cotton, and iodine. Abu Mahmoud had new types of cheese, which he set out on big plates and left in the care of Umm Mahmoud. He was wearing his new trousers embroidered with silver stripes, a new leather belt, and had a new headcloth fastened round his head. He looked like the *mitwalli* of the mosque, Haj Aziz, who stood near him. Between them were Abu Hashim, Abu Masoud, Abu Iman, Abu Ghanim, and Muhammad the builder. Blind Umm Aziz brought big holiday plates dotted with sweets, calling out, "Today everything is free for our boys."

They all came out: the coppersmiths, carpenters, ironworkers, and builders. Aunt Najiya's young voice parted the crowds of women and children before her: "Dears, clear the way a little for me. I'm ill and out of breath."

Bahija, La'iqa, and Aunt Naima raised their voices in prayer: "God bless the Prophet Muhammad. Protect them, O Lord, and let them be our protection."

Aunt Farida went up to the high roof and stood there. Her voice was inaudible and her face was indistinct, but she clapped and chanted. The front of the demonstration appeared at the intersection of the first houses passing down Great Imam Street, and she jumped and skipped, rushing amid the throng. Her clothing was white and her head was covered with a bright veil. She refused to wear her cloak on this day. She stood in front of Abu Mahmoud, who was holding the hands of the boys and girls of the neighbourhood, making us one circle.

Adil had not chosen a partner or a spot to stand in; he moved in our midst like a sleepwalker. Suturi and Nizar called out and laughed. Hashim was bursting with enthusiasm and played with his voice, wanting to release it. Mahmoud was far, far away, dripping with sweat. His voice erupted like a fit of dry, irritated coughing. "He's become a communist," the people said.

I did not understand what that word meant, though I had heard it as if my father had his pistol out and was chasing me. Mahmoud had changed; his face was harsh and his appearance was different, his luxuriant moustache was thinning, and a strange stillness had slowed his rapid gait. He had changed and become introverted; he was tense, no longer among us. When he stood near me, he used big words and the titles of thick books.

He came quietly and left secretly, and passed through the neighbourhood as we slept. The family's former sense of security was gone, and their easy kindness had become wariness. He was careful the way he looked at you, and when he locked eyes with you his eyes were like a threat. When you were with Firdous it was she he spoke to, and when you were alone you beat yourself in his name. Everyone in the neighbourhood bit their tongues and feared for him: grown-ups, family men, important people, all knew that "getting into politics means trouble," but they knew very little more and kept quiet.

Mahmoud began to disappear from school and home, from the neighbourhood and his neighbours, his close friends, and you.

The day he gave me a leaflet I was afraid, trembling and stammering. The first leaflet was like a first forbidden kiss. I could not move; my stomach was upside down and I nearly fainted. I knew that there was something like a bomb inside it, and if I touched it it would blow my hand and head off. I read it but only found Nasser's name mentioned once carelessly.

Muddled, I stopped reading and handed it back to him in silence. He vanished before me and left me only the temptation to read. I hardly understood anything; there was hardly anything I didn't understand.

We were all Nasserites. When my grandmother heard his voice she said, "I don't care if his nose is too big. But his voice – it's as though I've heard it before. It's like Abu Jamil's voice."

My father came into my room when he returned from Karbala, turned on the radio and tuned it to Voice of the Arabs, set down four glasses before him, clinking one against the other. He listened and said:

"Oh my God, deliver us by his hand from this black death."

Rasmiya kept Nasser's picture in her small work bag, and looked at him whenever she opened it; she kissed it and put it back in with the cotton and surgical spirit.

Umm Mahmoud, Abu Mahmoud; and blind Umm Aziz told us, "I wish I had my sight just so I could see him."

Farida was like quicksilver: when his name came up she paid attention and got excited. When anyone cursed him she kept quiet.

We pushed into the crowd, taken by his name and his picture, and when the electric current was cut off we spoke to him in the dark, Adil and I, in the name of the Prophet's household. Faces and bodies on roofs and behind windows cried out and threw sweets and nuts to us. Our voices rose as we saw him, young and warm. Then he went high up into the sky.

All Baghdad joined the insurrection that day. Cities, villages, and coffee-houses shut down, shops closed up, the universities wrote their banners and the students flew them, green, white, and red. The high schools let out most of their classes; each class covered the rear of the one before it, the faces of the police and their cudgels and sticks wanted even more of these bodies and heads. My father took the day off and came to Baghdad, telling his superior that his mother was ill. He took off his uniform and slipped into the crowd. I didn't notice him there, but Wafiqa saw him and smiled.

How often I had raced and skipped, run and strolled down Great Imam Street. I could see him now as he taught me how to fly. Nasser came and infiltrated our vocal cords and set all the secrets free. We entered into the rapture and began to chant: "Curse the English, curse reaction, down with colonialism and the Regent. Say Palestine is Arab. Down with Zionism." Stop stuttering. Fight. They fought with bare chests, necks small and large, and collars worn out from washing, and swore allegiance to him.

His voice was like piety, and my grandmother's shouted along with the rest: Down with the treaties. Down with tyranny. You memorized everything quickly. No one expressed his anger at the King of Iraq. Faisal II was absent from our cries, and stayed far from our voices. All the aunts and women of the neighbourhood loved him:

"He's a dear. He's still young. All the troubles have come from his uncle Abdulilah and the English."

The picture of the King of Iraq deceived young and old alike. He was handsome and sad, gloomy, yet fortunate as well.

Girls dreamed of him, and women worried about him. There was no blow aimed at him, and no wedding for him. Nasser took everything written on the banners and came to us. When he held court in Cairo, prisoners came out into prison yards and wrote his name with coal or their fingernails on the peeling walls.

When he gave a speech about the Suez Canal, the Arab radio stations divided homes into Nasserites and reactionaries, and the camp of those who did not know anything about it. Shops were closed, and homes were turned upside down in the search for a radio tuned to Voice of the Arabs or a leaflet slipped under old mats.

The neighbourhood was still in shock. Hubi's shop had been closed for seven months. We walked by it every morning and said "Good morning" to it. We passed by again in the afternoon and touched it with our hands. The dogs and cats licked the crevices of the place and congregated under the stone steps. Hubi was a bachelor, and his mother and sister cried for him; the whole neighbourhood was like a face engraved in acid. Abu Iman was carried home on the men's shoulders one night and everyone heard his shouting, his punches and curses. Rasmiya made nothing up. She bid a good morning to Abu Mahmoud, whose face was as sour as a squeezed lemon. His head was empty and his face was suddenly old. Mahmoud was nowhere to be found, in the neighbourhood or the school. When he passed it was after dark, and if he slowed his pace he did not offer a greeting. He moved to his uncle's house in the al-Fadl neighbourhood.

You still had Firdous, Adil, Nizar, Hashim and Suturi. You did not play with beads, or build mud houses; Mahmoud's top dug holes in one land and abandoned another. You imagined you were impossible to quantify, and that you would see him all through your life, inside the house and on the high roof. You opened your head and went down to the soul. You smoothed every road for him so he might settle there. You smiled as he helped you wipe away the handprints, your father's punches and the coarseness of the road. But after Mahmoud's death you brought him to the other side of the liver. I covered him with a bit of clean, thick cloth, tied him with the first laundry line, and established the place of residence preserved in the box of blood. The brightly coloured beads of childhood scattered – stolen, gilded with light touches, longings, and delegations of tears, and the spongy mud we mashed with our feet as we played sliding down slopes or streams. The first hours of the first meeting and pressure on the lock, and I hurried to hide him among the chapters of a year that would never return. When no one remembered him, I released him in the open deserts of the body. Mahmoud rarely came through that door; Firdous, too, fled from my grasp, raised her fist, which had grown, and loomed at my face, saying, "We're going to move near my uncle in al-Fadl, as well. The government is going to destroy this neighbourhood. My mother's looking for a new house for us."

You did not heed her words well. Your nose had not picked up the new scent. When you coughed, the dust and trash of your street was coughed up from your lungs. When you stood waiting for Adil at the gate of the school, you entered the race arena with him. Adil moved like a rabbit among the gaps in the lanes and alleys, free of the silence he had maintained since the night in the bath. He was divided on himself, and walked with every part of him wanting more division. When he left the school, he slipped away from me and raced to Khulud and her river banks. There he made the rounds of the street, the neighbourhood, and the school, the stones and the people. He stretched his legs out to the river's edge and played with his hands in the sand that breathed between his feet, fine and moist. He formed faces, numbers, and features. He looked at the Tigris and threw pebbles into the surface of the water, but did not look at Khulud's mansion. Looking at it did not do him much good, so he left the mansion to its creature and took her image into his soul. When the breeze blew over his chest, he wrote her name and flew it in the air, and when a wave reached him, it wet his rib that rended itself flesh and muscles. Everything he felt he threw away, and everything he threw away he expected to disappear. When I took him to my room he remained seated, and when I went out he stayed where he was. When we got something he ate, and when he was hungry, he did not utter a sound. I opened his books and read, and he read along with me, never making a mistake, never grumbling or complaining. He revolved around his only star, but never uttered her name.

That Khulud never went out and was never absent. She stayed in the airy halls of her mansion, going up to the high roof with its floor of many-coloured bricks, throwing him from afar an empty, folded piece of paper, with sticks of birds' nests, with a small, harmless pebble. She came down to the garden and sat on the wall, jumped and skipped, stuck her head out and disappeared among the rose bushes, emitting her golden laugh and throwing him all sorts of flowers we had never seen before. He scattered the roses on the stones, pebbles, and sand,

and they flew into the waves. Adil did not pick them up. He did not turn her way or greet her, nor lift a finger or bow his head. He only gazed at the opposite shore, at the fishermen and their old nets and corroded flat-bottomed boats while Khulud trampled his sand castles, obliterating his features and hopes. She walked behind him, a ribbed white ball in her hand, wearing a white dress and light sandals, her hair loose, with bright, new, yellow ribbons in her other hand. She was quick and boisterous, as pretty as a dove steeped in coquetry. Everything about her was petite: her round face, her eyes and nose, her eyebrows and delicate, blooming, laughing lips. She walked like a soldier, her steps sudden and movements brisk. She skipped after the ball, playing near the river, opposite Adil. She danced and leaped through the air. Her body undulated delicately; her skin was the colour of a flower, and her bones were fine, fed with vegetables and luxury. She fell near Adil and stood behind his back. He did not move. She bounced the ball, making small holes, touching him, passing the ball across his head. You were standing near both of them. You walked along, at a leisurely pace, not looking in their direction.

"My name is Khulud. What's yours?"

He did not turn around. "I know."

"What grade are you in?"

"I passed sixth."

"Me, too."

She took the ball in her hand and stood in front of him. For the first time he saw her. "Do you know how to braid hair?" She put the ribbons in his hand and turned her back to him. "Go on, braid my hair." There was nothing left of Adil but his arms. He turned to her and busied himself with her locks of hair, his hands burning with excitement as he felt them, separated and combed them with his fingers. He squinted, ornamented by sweat. His legs trembled and his hands shook. He spread her hair across her back, fondled it, brushed it forward, and got up hurriedly. Her voice stung him from behind his back, trapping him on the river bank.

"Afraid? Afraid?"

Adil ran, fell and got up again, his feet barely touching the sand, his arms fleeing her touch. The mud splattered on his clothes, his knees and hands. He raced, shouted, laughed, and did not disappear. The shore guided his steps. No one was there but the three of us. Khulud ran after him, and stopped and waited in the middle of the pebbles and stones, rolled and fell in front of him. She got up and her voice pierced the air: "Afraid! Afraid!"

She picked up pebbles and big stones and threw them in the air over him, at me, and at her house. Adil ran, remembering Munir's baldness and the blood on my hand from my wounds, and my aunt's spitting. He raced and buried his head in the wind. They raced one another; the first time they met each ran in a different direction. The beginning of this river is a drop of blood that moved and released its wailing in the circle of its area of the Mosque of Abu Hanifa as far as the dirt dam, as far as loud Khulud's light steps; Adil was as far away as the stars in Gemini.

I bowed my head. Khulud, in front of me, panted beneath Baghdad's bright sky and the Tigris waves crashing before us. I said, "He'll come back in a little while. Don't worry."

16

cの⊙の

Let us forget fear and put it aside, but it is present and tyrannous.

Only Farida beat it before her, and did not speak to it without mocking it. She approached her fear with natural muscles and found it work in the end: to make Munir stagger, with the rest looking on. My aunt remained the virgin, lifting up the title and contemplating it day and night. She took off the black dress, washed her dusty skin, and proceeded to put on a seductive nightgown; madness returned to her face.

She began to beat us, Adil and me, and Grandmother accepted it. She only wanted her: blameless.

Her voice sounded like a trumpet after months of long muteness. She went into the bath and wept there, shouted, and unleashed her voice upon us. She came out nearly naked, stood in the middle of the house, shouting, while Grandmother stood before her, praying and breathing on her, seizing and pulling her, encircling her with her arms:

"Dear, I have my voice back. Are you listening or not?"

She said: "Huda dear, Adouli, come and listen, dears, I'm afraid she'll go hoarse. I'm afraid so much talk and she'll lose her voice. Perhaps I should cut back a little and be like you, and talk little. What do you think, dears. Will I lose it?"

Farida changed. Her voice was a web of heavy-headed pins, and her silence too assaulted us. She took up all my father's weapons and plunged them into our flesh and our bodies, and we recalled Jamil, gasping. She beat us and we all cried, all four.

Grandmother held her head: "Please, Farida, my dear. Your voice has come back and it won't go away. God bless you, my dear, put your clothes on. I'm afraid you'll get ill."

"No, no, I want to go out, I want to walk in the streets and see the neighbours, take a walk, and sing, and say hello to everyone. I want to hear my voice again. I'm afraid of lies."

You stood far off.

"What's wrong with you? Are you listening to me? Don't worry, I won't beat you from now on. This is my voice. Tell me, Huda, have you gone mad?"

When she began to curse or laugh, when she insulted everyone, when she was cruel or talked nonsense, nothing could deflect her violence.

My father came several times, looking weary, sallow, and old. His clothes were faded and his shirt wrinkled, his boots dirty, and his face pale, melancholy, and unshaven, as if he had emerged from a shroud.

He did not shout or curse. He did not strike or torment us. *That* Jamil had been stolen for good, and we were even more frightened. When we were quiet he was uncomfortable, and when we went somewhere he vanished. When we stood in front of him he looked down at the ground, and entered into obedience to all things.

When he went into his first room in the middle of the house, he wandered. He handled the Qu'ran and stood a long time before it. He opened the closet and touched his best clothes, never looking in the mirror.

His eyes were lifeless, as if exhausted by hatred and rage. He did not take Adil in his arms, or call out to either of us. He was not tender, he was dejected and quiet. He now had other children whose entreaties and shouts he could hear: Saad, Raad, and Ali. He wanted a new gold-coloured star hanging on his shoulder to soothe his grief, so that he could move to Baghdad as an awesome captain. He continued to wait for that star for months and months, wild-eyed and menaced. Wafiqa deluged him with smiles and supplications, but he kept an even greater distance, and we were even more afraid. Grandmother sent for him and he did not come right away. When he went to Baghdad, he came and went at night, and he listened to Wafiqa's feeble voice:

"Listen well, Abu Adil. This is not the time for blame. Your sister must be divorced. The courts know her predicament, and we don't want scandals. Anything you say goes."

He did not raise his head or grumble.

"And so?"

"I sent you the claim the government issued on this house. They want to close off the street and the whole neighbourhood. I found a house in al-Salikh. Near your aunts' house, old but cheap – they were going to demolish it. The government will give us money, and you have to help us out a little, dear Jamouli. I know your situation; you'll get that new star, God willing, and hang it on your shoulder. Leave everything to me. What your mother says she does."

He raised his head to look at her, and said in a barely audible voice, "Is that true, Mama?"

"Your promotion was delayed for a year and a half. Your friends have become police captains. The world has changed and you have to change as well. Jamil, leave off drinking and swearing. How can you be promoted to captain when every day you're cursing the captain and the cabinet minister? You shout and you're quarrelsome."

She was quiet for a moment, watching him.

"Now get up and let's go and see the new house. As to later on – God will sort Farida out."

I did not hear the rest of what they said. I went out into the street and Adil followed me. For the first time I heard Adil's voice sounding coarse:

"It's true, Huda, we're going to be thrown out of this house."

I did not reply. We walked among the people hand in hand. Everything was in its place. The smell of cooking reached my nose, and I heard the voices of women as they dumped out dirty washing water in front of us. We walked along the muddy paths, through the mire and rubbish. Hashim rode an old bicycle, riding and falling. I turned away from him and he from me. We stopped in front of every bench, counting them and never making a mistake. We saw the holes in curtains and I hit them with my hand and moved on. All the residents chatted and exclaimed. We painted our names onto the metal electricity poles. The water pipes were rusty, and water leaked out of them.

The baker's shop was closed up. Abu Mahmoud sluggishly sold his wares, with crumbs of cheese scattered underneath the trays, and flies swarming over them. He did not shoo them away or cover the cheese with the palm branches which had yellowed and withered while the rest had fallen to the ground. He did not look up at Rasmiya, who was still limping from her beating. She plied her trade sticking needles in people's thighs and arms. We passed by her house, from which emanated the smell of surgical spirit and dried blood.

I saw Suturi and Nizar and bowed my head down. They watched in silence. Mahmoud was still absent.

There were still long queues in front of the shop of Hubi the butcher. He was back selling his wares looking vexed, neither singing nor joking. The pebbles and bricks of our grandfather's great house were strewn about.

I touched the walls of the houses, the gaps in the corners and the grains of dirt. I clung to the ample sand, and my dreams ran into the drains. We wailed, and the streets were changed beyond recognition by violence. We wept and comforted ourselves that all this outcry was warmth and that all this dust was roses.

"You're always quiet, Huda. Where shall we go? Now we're far from home."

"If you're afraid, go home. I want to go farther."

"No, I'm not afraid. But I want to cry and I can't."

Wafiqa said: "They used the last of their tears on the roof the day Iqbal died."

Blind Umm Aziz gathered her palm leaf tray, counted her coins, put them in a purse, and tucked it into her breast pocket. Abu Masoud the painter remembered that he had forgotten the light in his shop, opened the door and turned the light off, and turned to us. We looked at him. For the first time I saw him seeming dignified and handsome.

"Hello, uncle."

"Hello, my girl."

You wanted to throw yourself on his chest and sob into his shirt. The great lock on the door to his shop gleamed, and our eyes gleamed.

"Huda, where are we going? I'm not tired, just tell me where we're going."

Walk, Adil. Turn over the new visitors in your hands: doubt, remorse, and our friends who have stayed behind. Everyone walks, sleeps, closes their eyes, restores their bodies, and you alone are a traitor. Walk and don't be afraid of the muezzin's voice, or the stories of forgotten friendship. Don't sigh or move too quickly. Stay beautiful and quiet; stay mournful and afflicted, listen to Suturi's birds flapping their wings in this bloody sky. Spread your hands out and smell the sandstorm in the spring evenings as you fill your pockets with fragrant orange flowers, and throw them at Khulud's house. Smile Adil, if the stones gilded with dreams shout, if the concrete immersed in moaning lies, if the tiles laden with fever, fear and pain grow weary. Do not apologize for your nostalgia, Adil. Wafiqa once said that we were all sick. The books are sick. The table and love are sick. Do not bend or turn. Stay where you are. There is no time left in the world, and spending an hour here is uncanny. Laugh, Adil, at your wandering father, your absent mother, your proud grandmother, your diabolical aunt and your sister who did not love only you.

We stood in front of the dirt dam, pointing to the pessimistic Tigris and Khulud's house behind us. We looked at all the people there: You and Firdous will not meet again. She was the one to leave you. All those whom I loved left me, and all that time retraced its tracks to its original place. It went past my old dress and the ugliness of others and said, 'This far and no further. Do not turn back to pursue me, and do not look at me.' The dark mocking Tigris – I never do anything in front of it with ease. I was savage and cursed it with obscenities. But my thoughts turned to my father. I understood fatherhood, and instantly my father became precious. In our street, only my father was real. He never concocted stories or lied, never won or remembered.

Come and let me into your world. Give me the instruments with which you once beat me. Beat me, father, use electricity cables. Beat me, then sew up my wounds. Beat me and leave your marks on my flesh and face. Beat me and I will obey you a little.

We were attacked by pebbles and the fishermen's nets. The distant houses packed with lies attacked us, but they looked young and pretty. I learned lying early. I lied as easily as washing my face. Lying consumed me, and I memorized it. In the street we did not examine the truth or have time for it. We only told the truth when we were quarrelling, ill or had failed. When our truthfulness piled up, we agreed to wash it out of our mouths.

After six years or six months, take up the axes and chop up the flesh of memories. Do not shout or resist. Begin the parting now, but do not think of farewells.

They took us to the new house. You did not examine anything, not the guest room or the guests of this pain, not the little dead garden. You looked in silence and spat on the ground. The trees lined the street in a different pattern, "You will grow up anew here," Grandmother said.

"But I don't know anyone here to grow up with," Adil replied.

"Here things will be completely different," Grandmother said.

"But the fence is low," said Farida.

"We'll raise it," said my father.

Umm Mahmoud struggled like a fish.

"Our new house is bigger. Mahmoud will have his own room and so will Firdous. There will be room for guests and for new neighbours." She sighed, coughed, and added, "The boy's school is close by. He'll graduate from secondary school this year."

Firdous withdrew, becoming remote and cruel. She did not come or speak. You were the one who went to her. Your first parting was like your first meeting. You did not speak. You did not look at one another. You both fell silent. You did not touch. The suitcases were ready, and I could hardly recognize the house. Mahmoud's and Firdous's rooms seemed to me like a slaughterhouse. Everything was tied up, the beds and covers, the carpets, the kitchen utensils. Do not withdraw, do not cry, do not laugh. "Is it possible that I'll never see Firdous or hear her voice again?" I felt as sour as vinegar that had gone bad. I did not take a step or offer my hand. I did not want to see what was in front of me. I approached her and she stood before me, her head erect as if she had defeated me. I took her by the arms and shook her, but she did not shake. I bowed my head and looked at her legs. She was ready to fall down. I sensed that she was struggling to hide her emotions, then her gratitude towards me surged forward with her tears, without words. We tried to make the time pass quickly by filling it with small talk.

"My mother knows your new house and your aunt knows our new house."

"Give my regards to Mahmoud, but don't say any more to him than that."

My tears did not flow. They found a different way of expressing themselves, and they held themselves back.

I did not stay long. When they left, when they took their suitcases and dreams, when they took all the streets, those things would be the only things that had power over you. I slammed the door behind me and went out.

I went to everyone in his house and told Grandmother:

"We won't go until everyone else has gone."

Everything in our house was being packed. The chaos and confusion, and our very bone marrow. You go up to the roof and attack this universe. You put the legacy of the wedding into wooden boxes. You worked slowly, coughed, but did not cry. You looked at what was left in your hands. Anthills and cocoons, the trails of black and grey spiders, and dead locusts. There was no sign of Suturi's birds in the sky.

My father became effusive with his compassion: he became tender and indulgent. But my imagination had not killed his old cruel self, and my dreams had not conjured up such an honourable gentleman. He got his sister divorced from her cousin, and sorted out the new house. You had never known him to be so weak and in such a state. We feared for him more than before and our spirits were troubled.

Every week they took us to the new house. All the houses there were the same: two storeys, with bright exterior colours and sparkling windows. The children wore long trousers and clean shirts. All the girls walked confidently. I saw no lame children on my way, or any cross-eyed like Hashim. I did not see, on any of the fences of the houses, the title "nurse" scrawled in black coal, or stone steps. The entrances were roofed and the garages were spacious, the gardens were terraces with rose bushes and orange and tangerine trees. Each house was separated from its neighbour by fences painted white and light blue. From outside, the curtains looked very thick, and I could see no one behind them.

When we went home in the evening, we immediately went to bed. When everyone was quiet, I dreamed that I was walking. I turned on the taps and gathered up the soap in Baghdad to wash tongues and intestines. I forgot speech and swallowed its remnants. I shook, and stamped on the floor, and Mahmoud and I ate warm bread fresh from the oven. We divided it in half and watched each other fearlessly. When we saw the aeroplane in the sky, we laughed and smacked one another. Mahmoud thrust his face into mine as he said:

"When we grow up, Huda, we won't beat our children, and we won't pull their hair, and we won't make them run away to the shore in the afternoon. We'll go and swim with them. We'll ride the trains, and who knows? Maybe we'll ride in that aeroplane. Perhaps we won't see each other much. That's not important. I will see you when I grow up; I'll wait for your news from far away. Don't worry – I won't change."

I learned to write those expressions – I won't change; don't worry – every day of the week. Every hair on your head enters the race that is life. The runners tremble. The banners are wiped clean of writing: yellow, red and black. You run alone in the public squares. You do not listen to orders, you fall and you get up. You emerge from the crowd a zero, a fraction. Mahmoud was gone in the first round. He never said good morning or goodnight. Between the 'good' and the 'morning' came this wave of walking crowds. Do not ride it until the sand comes up. Do not befriend it until everyone joins you on top of it. Go in the opposite direction, and stop crying. What you are searching for you lose, and everything that you touch flies away. The neighbours lied to you, so you went to Rasmiya, Abu Masoud, Umm Suturi, Abu Hashim, and Umm Aziz. You went round that whole part of the neighbourhood. I went out into the vast square, skipped among the dirt and dry, fallen dates, and lifted my arms up to the date palms, felt the laughing tree and the beloved fronds, and brandished in my hand the bunches of golden fruit. I did not see anyone I knew. Everyone had gone far away. There was no weary advice or serious threat, no marvels erupting from the box of the world; no wonders poking their head out of ancient sacks. They left you no key and no wisdom to hang on your ears like earrings; on whose breast will you fling yourself? Who will dry your tears?

❧

Your grandmother and Farida getting ready, arranging things and measuring the height of the walls, the ceilings and the roofs. They shopped, changed things, sold, and managed, tired themselves out and came back more delighted. Jamil came off the train, not riding a car or falling off a horse; he comes as blessed as the corpse of a prince, and goes as pure as a hymn.

My grandmother told him:

"Jamouli, why don't you remarry? Leave Nuriya to her children and come here. There are a thousand girls who'd want you."

He did not look at her. It was as if he were breathing his last breath: "You mean Nuriya can't come into this house either?"

"You know that. Why do you torment yourself and me with you?"

"All that just for the late Iqbal?"

"And the children. Or did you forget your son?"

"No, I didn't forget. But Nuriya is pregnant now. Mama, shame on you and me. I won't divorce her."

"We'll look round for you first, and when you get the new star and get transferred to Baghdad and become a police captain, all the families will want you."

My father disappeared. Liquor incubated his torment. His house in al-A'dhamiyya was gone. He did not resist, or talk about it, or forgive. He was alone before his uniforms: the hated boots, the sad *sidara*, the silent pistol the olive-green colour of his uniform, and the prisoners' cells, all stung him.

Sometimes he visited them. He looked into the little peepholes at night and smiled. He called to them, one after the other. He got some names wrong but did not care. He poured it out before them and told them about the star he had been promised. No one knew what to say to him in reply. All that red dust, those gleaming pebbles and interrogations by night and silence by day flew before him as he tried to escape from the family's talk and the children's talk, and the unknown words which would lead him he knew not where. He got drunk and chattered and cursed, longing to be heard.

He needed a different mouth and tongue. Everything before him was silent and forbidden, dreadful and different. He knelt on the ground before the closed doors and wanted to eat the dirt. He patrolled the courtyard, his vision confused by the night. Was this Karbala or was it Iqbal's original sensuous voice and her cheap perfume?

It was his drunkenness driving his mother and his children, his wife and his sister, his illness and his temptation, and he slid down. He stood and probed his body and limbs. The savour of intoxication was strong, and his body was deranged, and smelled, and waited for the moments to come.

He had doubts about the stars as he gazed up at them in the sky, neither shining nor extinguished he scratched his throat, and groaned. He stood in the prison yard, repeating his children's names one after the other, and the name of the one living in his wife's belly.

Nuriya was gaunt, pale, and quick to flare up. She loved him and excited him. When she laughed, she looked at his body, which knew nothing but nightly arguments.

He told her: "If it's another boy we'll call him Najm – star."

"And if it's a girl?"

"I don't beget girls."

"But –

"Huda is a boy. She's not afraid of me or anyone else."

He looked ahead of him and sunk inwardly. Nuriya's body gave him vertigo. When he entered it he forgot everything except the star. He had not counted the columns and rooms of this courtyard. Why had he forgotten to? Its surface was like her thigh, and those eyes inside the peepholes followed him; their breathing, their sighs, and their silence. His legs tensed up. He wanted to piss on the ground. Even his urine sounded intoxicated. He walked and pissed, ran and pissed, not screaming or laughing. The sky appeared perforated to him; like Iqbal's sick chest. Nuriya and Iqbal. He raced as if the clouds were a silken bed, he flew through the air, the prisoners eyes followed him. He did not open the doors or move away from them. The *sidara* fell from his head, and he bent over to pick it up, and ran with it.

Alone, he dripped with sweat. No one came near him, neither Sergeant Jasim or Master Sergeant Sadiq. He was like a little star, alone and twinkling, which had slipped from the horizon and landed on a waistcoat.

Suddenly he began to scream. A long cry, a frightening snarl, and drawn-out sob. Alone, he ran, smiting his head with his hand, not seeing the wall in front of him. The walls had all been here. The prisoners had been here with him. Where had everyone gone?

He runs to the big faraway store room. He kicks in its door and lifts up the jerry cans of gasoline. He walks with them and puts them in front of him in the middle of the yard. He opens them, and a carnelian-red cataract gushes out. Within seconds, it vanishes into the ground, digging little holes that subdue the surrounding earth.

He dipped the *sidara* into the can and ignited it. He was working like a gravedigger. His hands went to work, undressing himself. His trousers were on the ground, in flames, and he laughed.

"That's for Adil."

The fire blazed and flared up into a fountain of light. His jacket was in his hand.

"This is for Iqbal."

He grasped the three stars in his hand. His hand was in flames. His fingers went into the fire as he pulled the stars from the shoulder of the jacket. He put them in his mouth. His wounds became unintelligible as the fire entered his mouth and burned his cheeks. He threw the gold stars high into the air, one after the other, and screamed:

"Take them! Give them to someone else! Take them and sell them at the public market. Take them and free me from their colour, shape, and weight. Take them, aren't you listening?"

He put his hand inside the can and stirred up the clothes with the *sidara*.

"They were too heavy on my shoulder. They were ugly in my neighbours' eyes. They were— "

He pulled his boots off and threw them into the rising flames: "And that is for the head of police."

The flames spread from the neck area down to the blazing sleeves. His undershirt was on the ground, but by the time he began to take off his long linen drawers, columns of men were running toward him. The staff sergeant and policemen clasped him from every side. They took off their clothes and covered him with them. They brought thick blankets and water hoses, and started to put out the flames burning his fingers and his hair. He laughed loudly and rhythmically and wept:

"I want the star. My mother lied. The captain lied. The star li— "

He wailed and wept. The men encircled him with their arms. They folded him as they would a garment, firmly grasping his arms, legs and body up to the neck.

He laughed as he was bundled off to Baghdad in a government car. Sergeant Jasim stood by his head, with Nuriya and her mother at his side. In their hands was a letter from the department: *dismissed from government service for health reasons.*

As we rode in the truck it seemed to me my father was driving. My aunt was in the new house. My grandmother sat beside the driver, and we swayed in the back. The sofa poked us with its wooden legs, and the new bride's boxes jostled us. We piled together, our feet seeking some footing among all the odds and ends. Our bodies cowered inside our clothes. Adil did not look back. I did not know anyone to wave to. Between the new house to which you moved and the ancient government hospital, the trail made by our blood stretched out like a ribbon that had just been unfurled.

Alia Mamdouh
❦

Alia Mamdouh was born in Baghdad in 1944. She graduated in 1971 from the al-Mustansiriya University and has been chief editor of *Al-Rasid* [The Register] magazine and *al Fikr al-Mu'asir* [Contemporary Thought]. She is still a regular contributor to the main newspapers and journals of the Arab World. Her first novel was published in 1973 and was followed by a collection of short stories. *Mothballs* was her second novel. Her third novel *al-Wala'* has just been published in Arabic. She currently lives in Paris.

Fadia Faqir
❦

Fadia Faqir was born in Jordan in 1956. She gained her BA in English Literature, MA in creative writing, and doctorate in critical and creative writing at Jordan University, Lancaster University and East Anglia University respectively. Her first novel, *Nisanit*, was published by Penguin in 1988 and her second novel, *Pillars of Salt*, is published in 1996. Fadia Faqir is Lecturer in Arabic langauge and literature at the Centre for Middle Eastern and Islamic Studies, Durham University. She is at present working on her third novel, *The Black Iris Crossing*.

Peter Theroux
❦

Peter Theroux was born in Boston in 1956 and educated at Harvard and the American University in Cairo. He has lived and travelled in Iraq, Syria, Lebanon and Saudi Arabia. He is the author of *The Strange Disappearance of Imam Moussa Sadr*, *Sandstorms*, and *Translating LA*, and the translator of several Arab novels. He currently lives and works in California.